The Service Technician's FIELD MANUAL

A Practical Guide for Pest Control Professionals

William H Robinson, Ph.D.
Technical Director, B&G Equipment Co.

EDITOR
Lisa Jo Lupo

GRAPHIC DESIGNER
Nichole Frye

Publisher: GIE Media, Inc.

Address all correspondence to GIE Media, Inc. at www.giemedia.com or call 800/456-0707.

Library of Congress Control Number: 2011939533

www.pctonline.com

PREFACE

I wrote this book for service technicians—the next generation.

Many of the topics covered are not in the standard workbooks, company manuals, or certification courses available today. But the next generation service technician will need to know and understand substrates, dose transfer, the link between application technology and application economics, and the basics of delivery systems. The methods and materials of professional pest control are changing. Technicians, trainers, and training have to match those changes. This book is a place to start.

The service technician is the day-to-day face of professional pest control. The customer base is constantly changing and with Internet access may be better informed or misinformed. Customers expect a trained individual who has the tools and experience to identify and solve their pest problems. I think the depth of the training currently available for service technicians is not enough to meet those expectations. If we expect more from the next generation of technicians, they will need more material and better trainers. Technical directors and supervisors are also an audience for this book; most of them came from the technician ranks and know what is needed and how to present it.

The topics presented here are grouped to make the book more useful as a reference text and training manual. It is not intended to be a read-through from chapter to chapter, and because of that, some information may be covered more than once. The topics are grouped to keep basic and applied material together for easy reference, and for use in training programs.

This book benefited from the help of many people and material from other disciplines. The two people most knowledgeable about application equipment and its repair are Cecil Patterson and Claude Thomas, and both have contributed to the book. Bob Heiney provided valuable insights to pest control and the role of service technicians. Skip Smith, with B&G Equipment Company, provided illustrations and valuable information on equipment design and mechanics. Some of the illustrations used are mine; others are adapted from various books and other publications.

Lisa Lupo and Nichole Frye took the content and concept of this book and produced an accessible-readable format; I sincerely thank their skills and vision. Dan Moreland put this team in place; his support is greatly appreciated.

William H Robinson
Technical Director, B&G Equipment Company
Jackson, Georgia

TABLE OF CONTENTS

CHAPTER 1

THE NEXT GENERATION SERVICE TECHNICIAN

Much is already expected of the Service Technician in professional pest control, but the next generation will be expected to know more and do more. The greatest change will be on-the-job decision making: What is the pest? What are the conditions? What control method(s) should be used, if any? What are the expected results? And, what does the customer want or expect to happen? All this must be processed quickly and applied correctly.

While Integrated Pest Management (IPM) was a good concept for professional pest control, in time it simply came to define what was already practiced or practical for most household pests. It is still useful, but not relevant. Pest control delivered by the next generation of service technicians will be situation specific—the individual technician will be able (and expected) to choose the insecticide, the delivery system, and the amount and location of whatever control agent is used. All these decisions will be made after arrival and based on the conditions and the customer.

Situational specific, an idea borrowed from Industrial Pest Management Specialist Bobby Corrigan's concept of rodent control, will define the next generation of pest control service and the service technician—as well it should. Service technicians are (or can be) trained and capable of making decisions; modern application equipment can deliver precise amounts; and modern insecticides are effective at concentrations as low as 0.01% and some rodenticides at 0.005%. The tools and the control materials are in place for the next generation of technicians to decide what to do with them.

INSECTICIDES

The finished concentration of insecticides has been getting lower and lower with each wave of new classes of chemicals. The first group of household insecticides had a distinct odor and dilutions of 1.0 to 1.5%, the low-odor pyrethroids crossed over the decimal point with dilutions of 0.1% and 0.2%, and now the no-odor diamides and neonicotinoids have gone another step lower with the added benefit of no toxicity to humans.

A major change for the current generation of service technicians is the emergence of the Reduced Risk insecticides, the class of chemicals that, because of their structure, have no toxicity to vertebrates. Indeed, the product labels on some new insecticides carry no signal word, such as "warning" or "caution." The next generation of technicians will know this to be standard for the insecticides they use: the concentration will be low, the efficacy will be high, the amount applied will be small, and the treatment sites will be few.

EQUIPMENT AND DELIVERY SYSTEMS

Without specifically designed application equipment in the hands of knowledgeable technicians, insecticides and rodenticides can not reach the pest to do their part. Modern insecticides will likely become more effective at lower concentrations and amounts, but

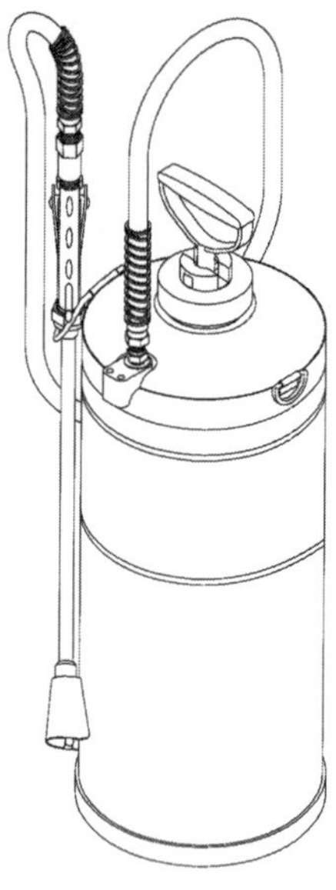

Compressed air sprayer

regardless of those qualities, they will depend on accurate delivery. To the residential and commercial customer, equipment is a visible signal of the technology they expect. Insecticides are odorless and applied to leave little or no visible residue, professional equipment is the first signal of efficacy.

The next generation service technician will have to grasp the reality of application economics. Designed into every sprayer, fogger, and bait gun is the capacity for a controlled delivery, and this must be clearly understood and utilized by the applicator. The costs of new insecticides will be justified by their efficacy with limited application, but only if this combination is realized by the technician. The business side of professionalism cannot wait for the next generation service technician to discover application economics; it will have to be taught.

CUSTOMERS

The next generation service technician will realize the influence that Internet access has on the relationship with customers. Various websites provide identification, information, and control methods, as well as the Material Safety Data Sheets (MSDS), and customers can have these before the first treatment.

All of this would be beneficial, except online identifications are often wrong, biological information misinterpreted, and control methods dated or slanted. The technician has to avoid these same pitfalls.

The next generation customer will invariably link successful pest control with knowledge, and knowledge with professionalism. This is an ideal customer. They are most likely not at home during treatment, yet are concerned about exposure to pets, plants, toddlers, and allergic reactions. The dry cleaner will tag a stain that would not come out, the car repair shop will detail the work and return the worn part, the next generation service technician will leave behind details on what was treated, what actions to avoid, and what to expect in the coming days.

TRAINING

For the generation of service technician that is now ending, training and testing was dominated by pest biology and habits—the number of eggs laid, their food preferences, their foraging distance, and how many died during laboratory evaluations. For the next generation, technology is much more important, in fact crucial, for the development of professional pest control. Development of sophisticated equipment and application methods will require more technical training and matching skills. This will not be filled by the current generation of trainers, training sessions, and materials.

Needed now are the materials, the trainers, and training sessions to match the ability with which the next generation of technician was born.

The long-accepted practice of on-the-job training for new technicians provides training on the art of professional pest control from an experienced mentor, but little of the science. Without more science and technology in the training, this industry will stall at the end of the runway.

The next generation of service technician has been raised amid technology and change, cursors and flash drives, and an app for everything. They will look for this in the workplace and accept the challenge—want the challenge.

Training economics will show that acquiring information costs time and money; on-site and off-site training has to show a return on investment for the individual and the company. For too long training has escaped the scrutiny of profitability; for the next generation technician the costs will be weighed. Companies as a whole and individual service technicians responsible for their time-profit margins will evaluate training opportunities in economic terms and will conduct or participate in training if the information is more specific, and not simply more, and as long as the economic benefit exceeds the cost of getting it.

TECHNOLOGY

Inspections, identifications, and applications by the next generation of service technician will be made or assisted by a small electronic hand device. Clipboards, graphs, and guessing will be replaced by electronic pictures, videos, and voice communication between technicians, managers, customers, and websites. On-the-job training will be more accurate, relevant, and repeatable, because it can include a short video from the website of the pesticide or equipment manufacturer. Similarly, on-the-job questions and identifications can be directly answered by the branch manager or technical director.

The performance of the next generation of technicians will be based on the collective strength of the technology used by the next generation of pest control companies.

Inspections, identifications, and applications by the next generation of service technician will be made or assisted by a small electronic hand device.

CHAPTER 2

IPM: PEST DETECTION AND MONITORING

Integrated pest management (IPM) programs are based on having information about the target pest, then making a decision on the control methods to use. Programs typically begin with collecting and identifying the pest, locating the infested harborages, and using chemical or non-chemical control methods. IPM programs continue on the basis of monitoring, with traps or direct observation.

INTEGRATED PEST MANAGEMENT

IPM began in agriculture; it provided an economic basis for decisions to use chemical control on crops. The objective of IPM is to provide an effective and economically efficient approach to pest control. It is based on the concept of reducing the number of pests to a level that does not cause an economic decrease in yield. This is pest management and not pest elimination. In pest management programs a low level of pests and damage are tolerated. IPM is a logical approach for production agriculture because it places pesticide application on a need basis and not on a routine-spray basis. IPM in agriculture also includes biological (non-chemical) methods when they are effective and economical.

The concept of reducing pesticide use and managing pest populations may be considered for household and structural pests. However, applying a traditional IPM philosophy of managing (not eliminating) pests such as cockroaches, fleas, bed bugs, termites, and rodents is not practical. The approach to pest infestations in schools, houses, hospitals, and food facilities must be elimination. Pests in the living environment can not be reasonably managed or suppressed to a low number. This simply would not be acceptable for health or aesthetic reasons. However, the integration of non-chemical methods into a pest control program can contribute to immediate and long-term solutions to pest problems.

IPM is sometimes considered to be the use of only non-chemical methods (no insecticides). However, this is rarely the basis of a successful approach to a pest problem, and ignores the characteristics of modern household insecticides. Many of the current products and formulations have no vertebrate toxicity, and have been granted Minimum Risk and Reduced Risk status by the U.S. Environmental Protection Agency (EPA). The predators, parasites, and other biological methods that can be successful in agriculture have little or limited effect on household pests.

INSPECTION

A thorough visual inspection of the site is the basic starting point for pest control programs. It requires knowledge of the target pest and experience with residential buildings, food plants, health care, and hospital facilities. Inspections require training and experience, and they take time. However, the time spent on inspection may result in less time in control and a more effective and cost efficient program. Inspections should include:

Inspection provides justification for chemical and non-chemical control methods, and a baseline for evaluating programs.

Environmental conditions. Consider all environmental features that can contribute to pest infestation or can influence prevention and control strategies:

- *Indoors.* Carpeting provides harborage for cat flea larvae and localizes their food after it drops from adult fleas feeding on house pets. The humidity of sink drains and garbage disposals attracts fruit flies, while the wet conditions and organic matter in bath drains provides the food for moth fly larvae.

- *Outdoors.* Vegetation surrounding buildings provides the food source for aphids and other insects feeding on plant sap, and the honeydew from these insects is food for many ant species invading buildings. Branches of shrubs and limbs of trees that contact buildings provide a bridge for carpenter ants to forage for food and nest sites. Poor rainwater drainage around buildings can provide favorable conditions for subterranean termites; moisture-damaged wood can provide favorable conditions for carpenter ants.

Distribution. Insect and rodent pests are not evenly distributed in the habitat; there are "focus points" where pest density is high. Inspections can reveal pest activity, and help to justify specific control methods.

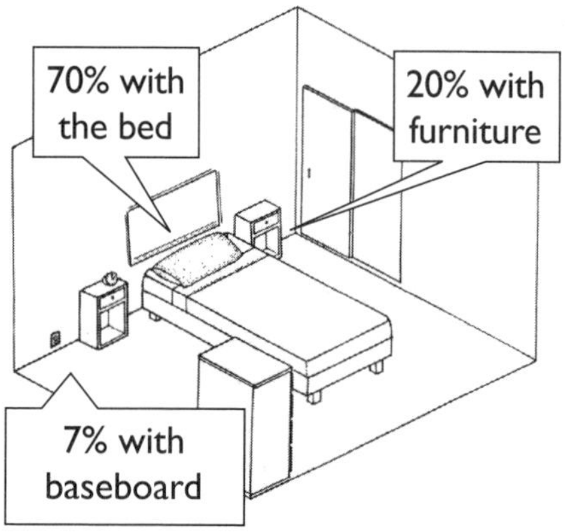

Bed bug distribution

- *Indoors.* Bed bugs infesting hotel rooms and residential bedrooms are most concentrated at the bed (70%), in bedside furniture (20%), and behind baseboards (7%) adjacent to the bed. Flea larvae in carpeting are concentrated where the household pet sleeps or rests.

- *Outdoors.* Overwintering boxelder bugs and Asian ladybird beetles gather on surfaces warmed by afternoon sun. Subterranean termite monitors are effective in soil where temperatures are consistent.

Sanitation. Poor sanitary conditions can decrease the effectiveness of pest control programs. Improved sanitation will remove some of the food and harborage that contributes to pest survival and limits control methods. Household clutter can seriously limit the effectiveness of inspections for bed bugs and reduce the residual activity of liquid and dust insecticides. German cockroach infestations are influenced by the availability of food and water, but mostly by water in the habitat. Females can live six weeks without food but only 12 days without water.

Documentation. A complete pre-treatment inspection provides the opportunity to document and report conditions. Attention can be given to sanitation problems and other features that must be corrected, or at least considered, in the control program.

Programs designed for food preparation and packaging facilities require carefully maintained written records. For these sensitive accounts, the initial and subsequent inspections must be recorded, along with the pesticides applied, the findings of monitoring traps, and the sanitation and structural improvements needed.

DETECTION AND MONITORING

Detection is typically done with traps that capture larvae, nymphs, or adults. The efficiency of the detection method, such as feeding blocks for rodents or sticky traps for insects, depends on placement. Each must be placed where it will be encountered by foraging or dispersing pests.

Trapping is commonly used to determine the presence of pests in the habitat prior to control, to monitor control effectiveness after treatment, or to estimate the level of a pest infestation and the need for treatment or retreatment.

Action Thresholds. Detection action thresholds serve as a component of IPM programs, with the goal of limiting the use of insecticides. This concept is based on initiating control when a specific number of pests is detected in the treatment area. In some IPM programs this number functions as a tolerance threshold or trigger for a control action. Thresholds may be based on the number of pests detected at one location or over time in several locations.

- *For household pests, the IPM concept of action thresholds has limited use.* The tolerance number may be as low as one, because of the potential for the development of a large infestation—as in the case of cockroaches, or the potential for personal injury—as with spiders or stinging insects. The detection of any subterranean termite activity at the perimeter of a structure is usually considered to be a potential threat. For bed bugs, the presence of a single insect or the experience of a single bite is the trigger to begin control.

For ***occasional or seasonal pests***, thresholds must be established within practical limits. It is impractical to expect that commercial and residential buildings will be free of all insects and spiders.

Monitoring: Post-Treatment. Traps can help in the detection of infested harborages before treatment and serve as a monitor of pest activity after treatment. The goal of monitoring in pest management is to document changes in the pest infestation following a control action, and to signal the need for further action.

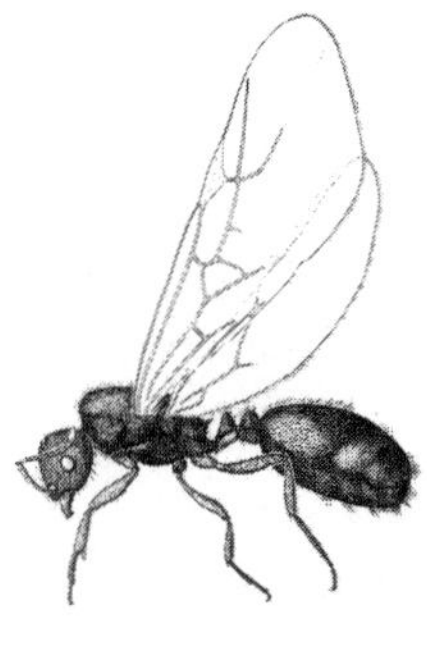

Winged adult ant

HOUSEHOLD AND STRUCTURAL PESTS

Ants. Ant infestations are often indicated by the presence of winged adults indoors. Infestations can be monitored by using feeding stations baited with a non-toxic food, then, at regular intervals, counting the number of ants visiting each station. Checking monitoring stations at the same time each day can provide information on abundance and colony activity.

The action threshold for ant control is determined by the tolerance level of the residents of a site. In IPM programs for schools, the action threshold for ants in classrooms and kitchens may be as low as one or two. Post-treatment monitoring is usually done by counting the number of complaints from building inhabitants, without regard to the ant species involved.

Most ant control programs start after a report of ants indoors. However, ants in buildings may also be non-infesting species that collect honeydew from insects on plants adjacent to the building.

Bed Bugs. Bed bug infestations can be detected and monitored with:

- *Specially designed traps* which typically utilize a combination of a pheromone lure, heat, and CO_2 to attract adults and nymphs from the immediate area. The insects may be held by a glue board or a pit-fall feature of the trap. The traps serve to confirm the efficacy of control methods and monitor for re-infestations. They are effective in locations where bed bugs are actively seeking a

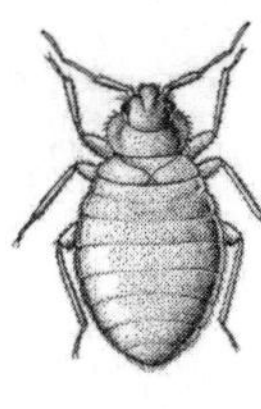

Bed bug

Sticky traps designed with 30° ramps and sides can be used to capture bed bugs. A ramp-based sticky trap acts as a pit-fall trap, with bed bugs falling onto the sticky surface when they crawl down or fall from the ramp.

blood meal. Interceptor traps for bed bugs capture those that attempt to crawl up or down furniture legs.

Inspection of a typical hotel room may take a skilled technician about 20 minutes, while an inspection by a detection dog may require 2 to 3 minutes.

- *Scent detection dogs* have also been used effectively for inspection and detection of bed bugs. A qualified dog is able to identify small numbers of live bed bugs, and can discriminate live bed bugs and viable eggs from empty egg shells and cast skins from a previous infestation. As with other pest-detection tools, scent detection has limitations and is not always definitive. Scent dogs rely on what they can smell, but sometimes the odor from bed bugs in the location is not available to the dog. The reasons for this usually include location of the bugs, air flow in the room, and room temperature.

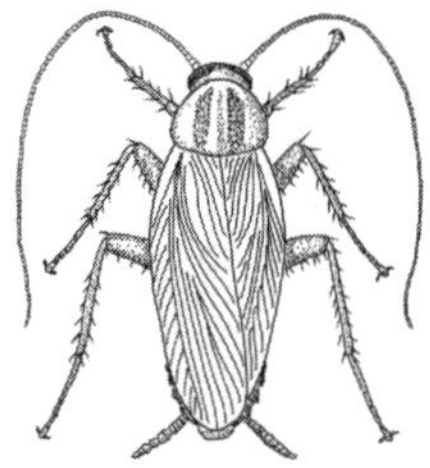

Cockroach

Cockroaches. Cockroach infestations can be monitored with sticky traps. Rectangular traps with a low profile can be placed under cabinets and kitchen appliances—out of public view but close to potential harborages. A 30°-angle entrance ramp to the sticky surface significantly increases trap catch. Fewer cockroaches escape these traps, and they are capable of securing adults and nymphs.

It is a common misconception that the infested harborage is in the opposite direction the captured cockroaches are facing in the trap. This assumes that cockroaches walked in a straight line from the harborage to the trap, which is unlikely.

It is recommended that traps be placed flush against surfaces, as those placed just 3/4 inch from walls catch about half as many cockroaches as those placed flush against them. Monitoring in residential and commercial facilities is usually done with two to five traps, which samples about 5% of the infestation in each 24-hour interval.

Dust Mites. The presence and level of infestation of dust mites can be determined by several methods, including vacuuming sections of carpeting or using a lint roller on floor surfaces. The presence of dust mites can sensitize humans and cause asthma, rhinitis, and dermatitis.

Action thresholds for programs to reduce mite allergens are difficult to establish. There is a range of human sensitivity to dust mite allergens, and controls may be short lived. Mites can re-infest from other locations, including office buildings, automobile seat covers, and clothing. Dust mite populations in summer can be as much as seven times greater than in winter. To prevent asthmatic symptoms or relieve an existing response to dust mites, the threshold limit is 100 mites per gram of dust.

Flies. The most troublesome indoor fly pests are the red- and dark-eyed fruit flies, and phorid flies. Locating the source or breeding site for these small flies can be achieved by considering the habits of the adults and larvae.

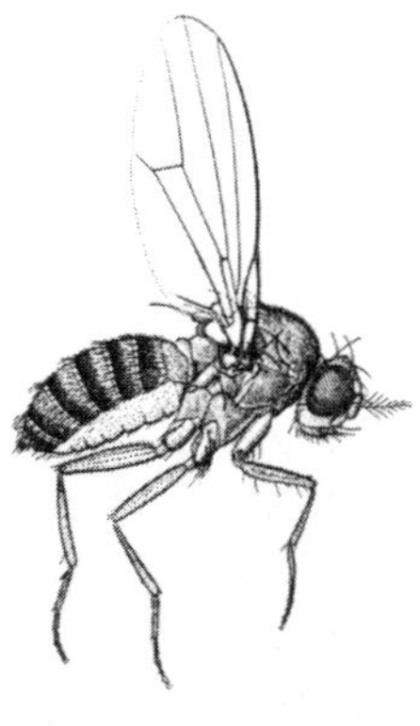

Fruit fly

- *Red- and dark-eyed fruit flies.* The red-eyed fruit fly is usually found in residential and restaurant locations; it breeds in decaying fruit and garbage. This is the species that hovers over alcohol drinks and humid sinks. The dark-eyed fruit fly is found in commercial kitchens and breeds in rich, decaying organic matter. It rests on walls and ceilings and does not hover over food. Detecting and eliminating the breeding site will eliminate ongoing infestations.

Female fruit flies prefer to be in humid locations, such as sinks, and areas with vegetable and fruit odors, such as garbage, but these are usually not their primary breeding sites.

- *Phorid flies.* These small flies are associated with sewage and other organically rich substrates. Their appearance in large numbers is usually an indication of a break or failure in a drain pipe. Flies coming through floor drains may be an indication that there is a break in the sewer line. When there is an infestation in a septic tank, there may be large numbers of adult flies in bathrooms.

Phorid fly

Phorid flies are distinguished by the shape of their legs and wings: the legs are long and slender, and the wings have few veins.

Termites. Detection of termite infestations relies on visual inspections. For drywood termites this may be kick-out holes and granular frass. For subterranean termites it may be mud tubes on masonry walls or live termites in ground detectors.

- *Ground detectors.* Termites around a structure are usually detected using wood stakes or other cellulose food sources placed in the ground. The length of the wood stake or plastic station is designed to intercept termite species that tunnel close to the surface and species that tunnel deeper in the soil. Once the food source is encountered by foraging termites, others from the colony will be recruited. Periodic inspection of monitors will reveal termites in the above-ground food source.

  In-ground detectors can be installed any time of the year. Subterranean termites often remain active during winter and may feed in moist soil where the temperature is 40°F or higher. However, termites are more active in spring and fall, when soil moisture and temperatures are high. Soil in sheltered and shaded locations around buildings can have extended periods of moisture, and it may also be protected from seasonal temperature extremes. In open areas, soil moisture is usually low, especially in summer. Dry, hot soil is not a favorable foraging site for subterranean termites.

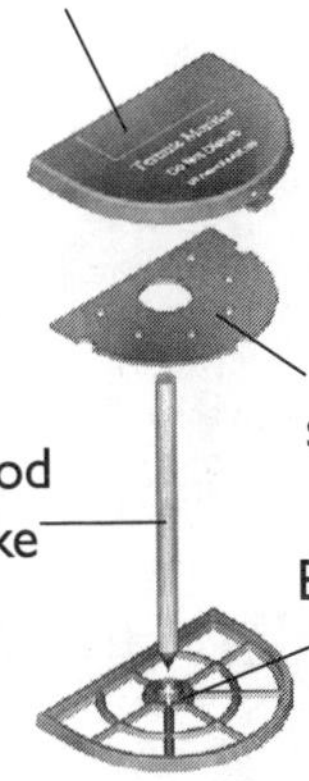

Ground detector

Detectors can be positioned at the edge of the liquid-treatment zone, where they can monitor the building perimeter and discourage disruption of treated soil.

Termite control programs typically include installation and maintenance of detection stations around and within crawlspaces of treated structures and at perimeter locations.

- *Swarmers.* The emergence of a large number of winged swarmer termites inside a structure or around the outside is an indication of an infestation. A swarm indicates the presence of a colony that has been developing for many years. In general, the number of swarmers is proportional to the size of the colony. Most subterranean termite colonies produce swarmers by the time they are five years old, but large swarms usually don't occur until the colony is eight to 10 years old. Formosan termite colonies may not produce swarmers until the colony is about 15 years old. Drywood termite colonies produce swarmers after about three years of development.

- *Electronic detectors.* Electronic tools can detect the elevated temperature and humidity that indicate the presence of termites and carpenter ants. Detection radar devices show the extent of activity and the direction from which the termites traveled to get to a particular point. This may permit detection of the entry point and the source of the infestation.

  Thermal imaging cameras detect heat radiation. They process the heat of infrared light and display it on a screen. The higher the object's temperature, the brighter the

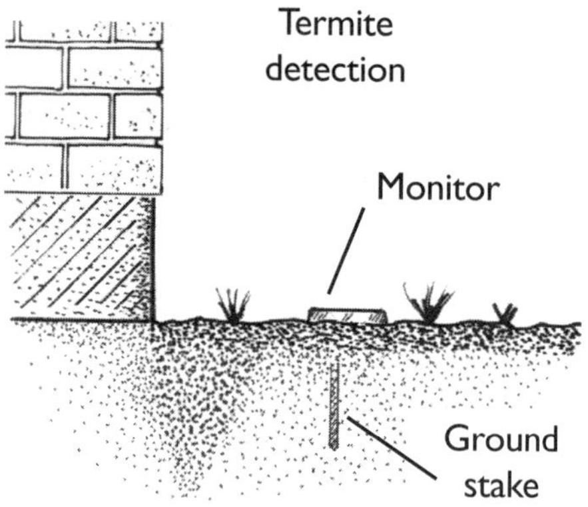

image is on a screen. These cameras can locate termite risk areas but may not confirm the presence of small numbers of termites.

- *Moisture meters.* Wood moisture content is useful in establishing the presence of conditions suitable for decay fungi and subterranean termites. The resistance-type moisture meter measures the electrical resistance between two pins inserted into the wood. These meters are reliable in the moisture content range of 7% to 27%. The number indicated on the meter scale represents the moisture content of the piece to the depth of the pin electrodes. Non-destructive moisture meters (so called because they do not insert pins in the wood) are based on measurements of the impedance of low-frequency signals transmitted into the wood.

- *Detection dogs.* Termite-detection dogs can locate active infestations of subterranean termites. Dogs can detect small numbers of termites in wood behind walls and in inaccessible areas. Their effectiveness as detection tools has limits, but the combination of a dog and an experienced technician is an effective and commonly used strategy.

- *Fluorescent markers.* Termiticide dust formulations can have the added feature of fluorescence. Termite workers that have eaten the cellulose bait can be detected by a yellowish glow using ultraviolet light. This permits tracking of termites that have been exposed to the dust, and possibly determining if those found in close proximity to the treated structure are from the same colony as termites previously exposed to the dust.

Wood-infesting Beetles. Wood-infesting beetles can be detected by:

Adult beetles emerge from May to early July, so marking the surface of infested areas to determine activity should be done in late fall or winter.

- *Exit holes in the surface of wood.* However, the presence of holes, or even frass coming from holes, is not positive evidence of an active infestation. Exit holes persist long after all the larvae complete development or have been killed with an insecticide treatment. Frass may be jarred from feeding tunnels and fall from emergence holes. Covering the wood surface with a thin coat of paint will identify the old and reveal new emergence holes produced the following spring. In areas with only a few emergence holes, each can be individually marked.

- *Chewing sounds.* Full-grown, old-house-borer larvae can be heard feeding in sections of structural wood, such as wall-framing studs or floor or attic joists. However, this sound can also be confused with scratching or gnawing sounds made by mice. Several longhorned beetle species infest pine firewood, and their feeding can also be loud and distinctly heard.

Powderpost beetle larvae can not be heard feeding in wood; these larvae are too small to produce an audible sound.

- *Structural damage.* Establishing an action threshold for anobiid powderpost beetles in pine framing is difficult. Infestations that result in as many as 70 emergence holes per square foot do not significantly weaken the wood. At this density of emergence holes, the modulus of rupture can be reduced 31% to 37% and the modulus of elasticity reduced only 21% to 27%. Compensation is made for these strength reductions by load sharing on complete floor systems.

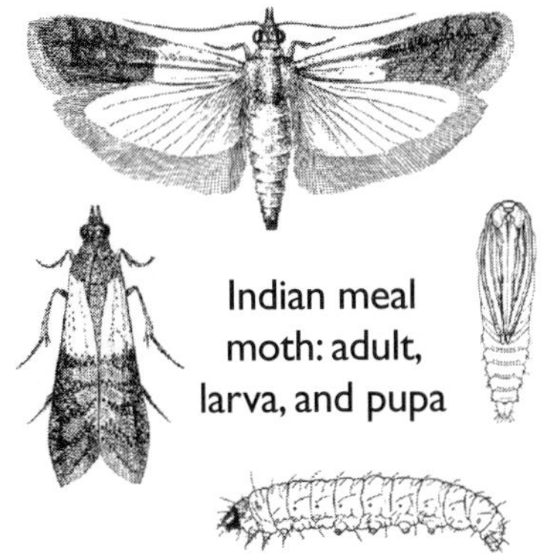

Indian meal moth: adult, larva, and pupa

STORED FOOD PESTS

Moth and beetle pests of stored food can be monitored with sticky traps that use a pheromone as an attractant. Most traps target a single species, but some use pheromones formulated to attract several closely related species. Pheromone traps can provide information on the species involved, seasonal prevalence, locations infested, and the presence of the adult stage in the population. Pheromone traps release a female sex pheromone or a male aggregation pheromone that attracts only males.

- *Placement.* The volatile attractant in pheromone traps must be widely distributed in the habitat. The actual number of traps used is based on achievement of uniform and three-dimensional coverage in the space. Traps can be suspended 6 to 12 feet in the air and spaced 30 to 50 feet apart to ensure maximum coverage of the pheromone. Traps should not be placed near outside doors or windows to prevent attracting insects from outdoor populations. Traps placed adjacent to walls often capture more moths than free-hanging traps.

The action threshold for stored food pests depends on the facility and the food product. For some it may be zero tolerance; for others zero pest tolerance is not needed but zero insecticide is required.

- *Infestation location.* Sex pheromone traps can be used to locate a pest infestation source in food warehouses and other large buildings. The basic strategy is to establish traps on a large grid pattern, then systematically reposition the traps based on the number of insects captured. Set pheromone traps in a grid pattern throughout the space. After one to two weeks, reduce the number of traps in the grid pattern to the regions that have the highest capture number. Allow two to four weeks for the traps to continue trapping. Reduce the grid pattern of pheromone traps by concentrating on the traps that have captured the largest number of pests; this will pinpoint the location of the breeding site(s).

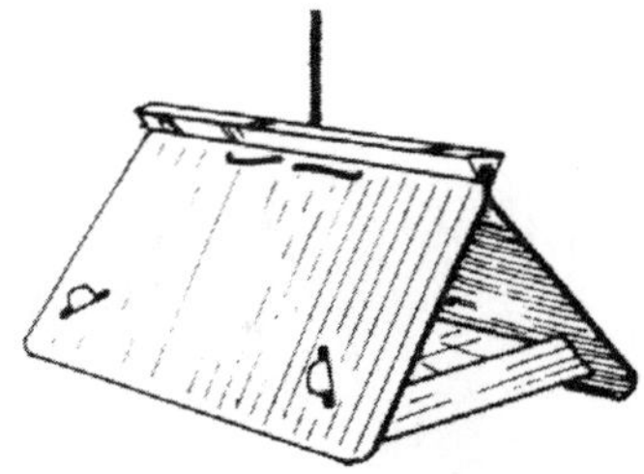

Pheromone sticky trap

- *Information.* When adult capture fluctuates between high and low numbers, it can indicate that a portion of the pest population is in the larval or pupal stage. Thus, control methods targeting the immature stage can use the adult capture numbers to properly time application.

BITING AND STINGING PESTS

Mosquitoes. Mosquito populations are monitored with light- and carbon-dioxide-based traps. Complaints by local residents of biting mosquitoes can also provide information on abundance, and help to direct placement of traps for more precise information. An ongoing monitoring program for adult mosquitoes provides information on local species, seasonal abundance, population dynamics, and results of control operations.

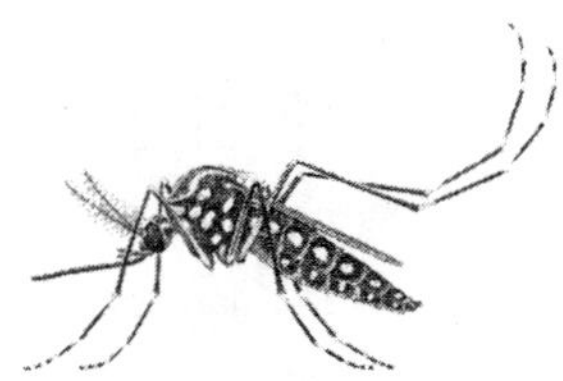

Mosquito

- *Adult traps.* Carbon dioxide (CO_2) traps are used for large mosquito control programs. They operate as suction traps baited with dry ice as the source of CO_2. Female mosquitoes are lured close to the trap by the CO_2, and the suction draws them in. However, adult sampling is limited to those species of mosquitoes that are attracted to light traps.

- *Ovitraps.* These traps are designed to attract females seeking an egg-laying (oviposition) site. Ovitraps consist of a small container that holds water and has an

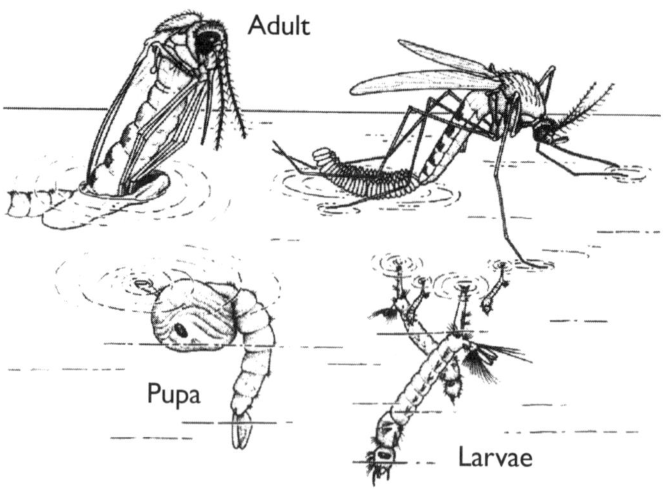

Life cycle of a mosquito

overflow hole in the side near the top rim. When used as a lethal trap, it includes a strip of paper or plastic treated with insecticide.

- *Aquatic sampling.* Determining larval breeding sites can be done early in the mosquito breeding season before adults become common. Sampling is done with a dipper, typically comprising a small white pan with a long handle. The pan is gently dipped into the water so the surface layer runs into it. The collected water is examined for the presence of mosquito larvae and pupae. The best sites to sample are water margins and areas within emergent vegetation.

  High numbers of large larvae indicate that it may be several days to a week before adults emerge; high numbers of pupae indicate that adults will be emerging within one to two days; high numbers of small larvae indicate that it may be five to 14 days before adults emerge; and, pupal skins floating on the surface indicate that adults have already emerged.

  Tolerance levels for mosquitoes are related to the density of adults in a residential area. Action thresholds have been based on the number of female mosquitoes caught per night in traps placed around the mosquito control area. The original threshold number of adult females was based on four bites per night as exceeding the tolerance level for mosquito bites.

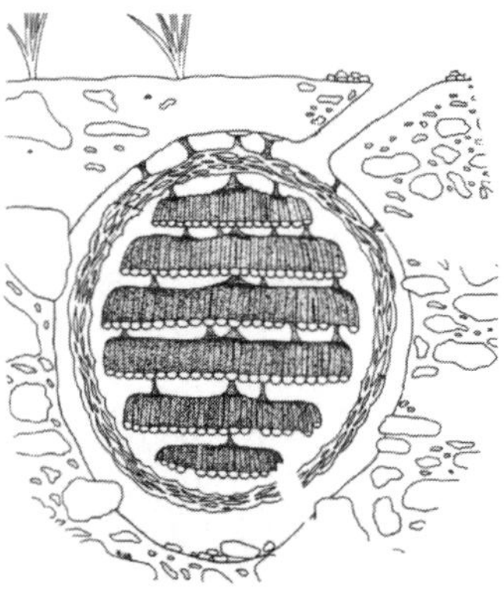

Yellowjacket nest

Yellowjackets. Commercial traps can be positioned in public-use areas to detect the presence of foraging yellowjackets in summer and fall. The tolerance level for yellowjackets is subjective; these are stinging insects and it may be as low as one. Ground nests create problems when they occur around buildings and in recreational areas. Urban residents generally consider one to four yellowjackets near a house to be a safety threat.

Scorpions. Scorpions commonly occur around the perimeter of houses and other buildings, as well as indoors, in many regions. There is no easy way to keep them out of houses, and no easy way to inspect or detect their presence once they are inside. Ultraviolet flashlights can be used to inspect for these pests. Scorpions glow a yellowish green under blacklight. The dorsal and ventral plates on the body usually show the fluorescence response.

Rodents. Monitoring for and detecting rodents can include:

- *Monitoring blocks.* These blocks lack rodenticide, but contain other ingredients that make them attractive to rodents. They are typically placed in a rodent bait station. Tooth marks from rodent chewing or complete consumption of the blocks is an indication of activity in the area. Once rodent presence is confirmed, these blocks can be replaced by those with an active ingredient. Some nontoxic blocks include additives, such as a pigment that colors fecal droppings or makes them respond to ultraviolet light. Tracking dust works on the same principle: a pigment is added to a non-toxic dust that is carried on the feet, tail, and body of the rodent.

- *Ultraviolet-light (UV) inspection.* The urine of rats and mice responds to UV light, and this feature can be used for detection and monitoring. Rats and mice defecate while walking and running; their paths and runways are littered with fecal droppings and covered with urine. UV lights can be used to identify frequently used travel routes and determine the extent of an infestation. Rodent urine is deposited in a series of droplets along a line. Norway rats and mice may drag their tails behind them, smearing the lines of droplets. Rodent urine will have a bluish-white color when relatively fresh, and a yellowish-white color when several days old.

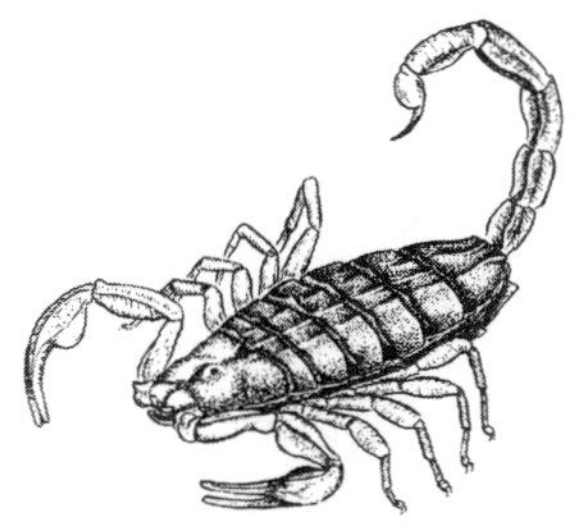

Scorpion

Other materials that glow under UV light include bleach, glue (on cardboard boxes), refrigerants, oil, and hydraulic fluids.

CHAPTER 3

NON-CHEMICAL CONTROL METHODS

The control of any pest depends on the elimination, modification, or contamination of one or more of its requirements for life. These include food, water, air, harborage, favorable temperature and humidity, and reproduction. Non-chemical control strategies include a range of pesticide alternatives or indirect methods that affect one or more of these requirements. While some may be practical and effective when used alone, they are usually most effective when used within an overall pest management program. Non-chemical approaches to pest control include biological agents, such as parasites and predators, and mechanical and physical measures that target the pest habitat or resources.

BIOLOGICAL AGENTS

Various arthropod predators and parasites attack household insect pests. Some are general predators, such as house centipedes and wolf spiders, while others are more specific, such as parasitic wasps. Although they can be very effective in killing their host, they rarely reduce pest infestations to an acceptable level. In addition, indoor use against household pests is not likely to be acceptable or practical. Microbial pathogens, such as viruses, bacteria, and fungi, are species specific and are only effective under ideal conditions. Most biological control agents are relatively safe in the human environment; they are simply not practical.

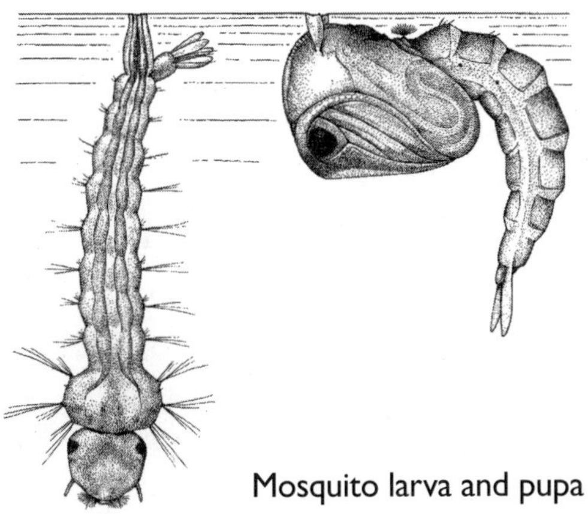

Mosquito larva and pupa

Predators and Parasites. House centipedes and wolf spiders are predators of indoor insects. However, their success is limited, and they are generally not welcomed by homeowners. Several species of small parasitic wasps attack household and structural pests, but they rarely reduce pest infestations to an acceptable level, and the general public has limited tolerance for additional insects indoors.

Some species of parasitic wasps attack the egg cases of cockroaches. Female wasps lay their eggs in the egg cases and the resulting wasp larvae destroy the eggs. The ensign wasp, *Aprostocetus hagenowii,* attacks the egg cases of American, smokybrown, and oriental cockroaches. This small wasp can occur indoors, but it is not a threat to humans.

Bacteria. The bacteria, *Bacillus thuringiensis israelensis* (Bti) and *B. sphaericus* (Bsph), kill the larval stage of mosquitoes and black flies (Simuliidae). Bacteria cause the fatal milky disease of Japanese beetles that attack turfgrass. Grubs ingest the bacteria spores along with soil and roots, and the spores germinate in the insect's gut and kill it.

The biological or microbial cleaners used by pest control professionals are concentrated mixtures of bacteria. They are applied to indoor drains in liquid or foam formulations to prevent or remove the build-up of organic debris. The bacteria in the dilution multiply rapidly, and their large numbers remove the organic matter—the insect food source—from the substrate of the floor drain or pipe. The bacteria penetrate the surface of the organic substrate and secrete enzymes that slowly digest it; some bacteria produce reproductive spores that remain alive for long periods, growing whenever food is available. Biological cleaners are typically used to control dark-eyed fruit flies and moth flies that breed in floor drains. The bacteria consume the substrate that serves as the food for the fly larvae.

The residual activity of biological cleaners is limited, because some bacteria produce methane gas when digesting organic matter, and the gas is toxic to the bacteria.

Bacteria used in biological cleaners occur naturally, and their use presents no threat to humans or food products.

Ensign wasp

Fungi. Fungi invade insects by going through the cuticle and attacking the internal organs. Strains of *Metarhizium anisopliae* have been tested to control subterranean termites. The major factors limiting efficacy are the behavioral responses of healthy termites to the spores, and behavior of nest mates infected with the fungus. Exposure to the insecticide imidacloprid can lead to an increased fungal infestation and faster colony collapse than either the fungus or the chemical could achieve on its own.

PHYSICAL BARRIERS

Physical alteration of urban structures or features can reduce or prevent access by pest arthropods or limit their harborage and breeding sites. Screens, caulking, impenetrable brushes at the bottom of doors, and air curtains have been used successfully in household pest control.

Bed Encasements and Covers. Encasing a box spring and/or mattress within a plastic or insect-proof material is effective in limiting bed bug infestations and exposure to dust mite allergens. In the case of bed bugs, the covers remove harborage and egg-laying sites; in the case of dust mites the covers act as a barrier for airborne allergens.

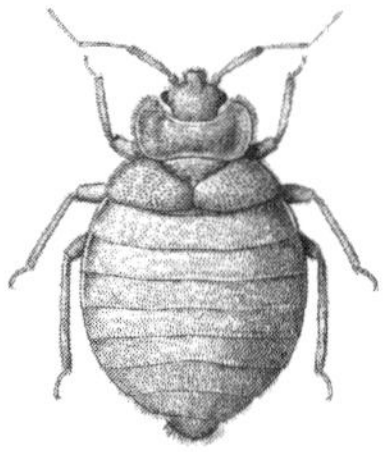

Female bed bug

- *Bed bugs.* Mattress and box spring covers and encasements used in bed bug elimination programs are designed to be bed bug bite-proof and escape-proof. These covers create a completely sealed enclosure to prevent any movement of bed bug adults and nymphs into and out of the mattress or box spring. The cover prevents female bed bugs from laying eggs on the mattress seams and the box-spring frame.

  Encasements designed for dust mite or other allergy protection may not have the special seams and zippers necessary to deter bed bugs. Covers intended for bed bug elimination have specially designed zipper end stops to prevent bed bugs from escaping.

- *Dust mites.* In dust mite management programs the mattress is fully enclosed. The cover blocks mite allergens and the passage of skin scales into the mattress. To do so, the material must block particles as small as 0.2 microns, thus non-woven material made entirely of polyethylene or polypropylene is commonly used for dust mite covers.

Encasing mattresses and pillows can significantly reduce the allergenic mite feces, although mites will survive. This strategy is usually combined with hot washing of bedding to further reduce allergen levels.

Shields, Sand, and Mesh. Sand and gravel aggregates, impermeable membranes, and metal mesh are used to physically block termites. They are placed around the perimeter or underside of a building as a permanent, non-chemical protection zone. In general, their use is limited to pre-construction application. Professional installation

is necessary and the procedures are critical to the long-term success of the barriers. Once in place, they are unaffected by soil type, temperature, or moisture.

- *Metal shields.* Metal sheets can be placed between the top of the foundation wall and the sill plate for subterranean termite protection. The metal sheet is continuous across the entire top of the wall, then extends downward another 2 inches at a 45-degree angle.

The purpose of the shield is to force termites to come to the surface of the foundation wall if they are tunneling within the blocks, or to interrupt their path if shelter tubes are built on the wall. Termites are not able to construct shelter tubes across the 45-degree angle, so when they reach a shield, they are forced to build the tube around the shield, where they can be detected during inspections.

Termite shields are usually considered to be a barrier against subterranean termites, but they are actually a means of exposing termite tunnels and improving inspection.

- *Barrier sand.* Termites are unable to move particles larger than 1.0 mm with their mouthparts, but smaller particles can be carried away by workers. On the other hand, particles that are 3 mm or larger provide spaces that termites can crawl through. Thus, to block termites, the effective size range for barrier sand is 1.0 to 3.0 mm, or 16-mesh, coarse sand. The sand can be used as a barrier around foundations, perimeter trenches, and crawlspaces; inside masonry block voids; and around retaining walls. Certain grades of sand-blasting sand and crushed volcanic cinder can be used for barriers.

- *Impenetrable mesh.* Mesh which is used as a barrier is stainless steel and corrosion resistant with spaces too small for termites to penetrate. It is used primarily with concrete slab-on-grade construction, where it is placed on the soil or aggregate bed before the concrete is poured. Materials available today have the flexibility to be formed around all construction features, and tests of the stainless steel have shown no significant corrosion after 16 years of in-ground service.

Caulking and Structural Modifications. Structural modifications to building exteriors may provide only limited control of crawling pests. Screening crawlspace vents, sealing cracks and spaces around plumping pipes and electrical conduits, and filling weep holes in masonry walls incur labor costs, but may not reduce pest infestations in the short or long term.

- *Cockroaches.* German and brown-banded cockroaches utilize the narrow spaces along edges and within cabinets and counters in bathrooms and kitchens. German cockroaches utilize narrow openings as harborages, and removing these from the habitat is virtually impossible. The first-stage nymphs prefer harborages of about 0.5 mm wide (the thickness of a dime), and adults prefer harborages about 4.8 mm wide (the thickness of 4 dimes).

  Thus, caulking cracks and crevices in kitchen and bathrooms removes only 41% of the total harborages that can be sealed; the remaining 59% remain open and available to cockroaches.

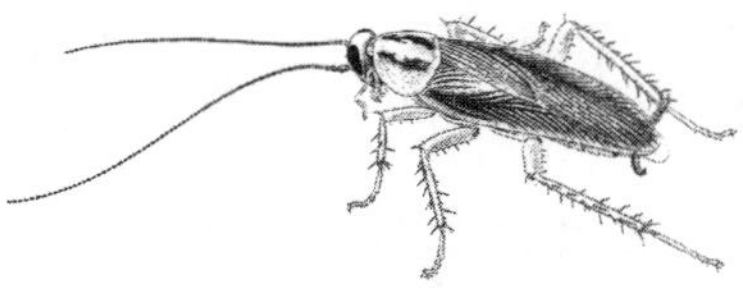

German cockroach

Caulking results in providing cockroaches a more secure harborage while preventing application of insecticides.

American and oriental cockroaches harbor in natural openings around buildings and the structural features of masonry wall, concrete steps, and foundations. Making structural modifications to the outside of buildings does not improve control, as modifications may reduce access to only a small percentage of all potential harborages.

- *Asian ladybird beetles, boxelder bugs, and stink bugs.* Pest management programs for these pests usually include spraying of a residual insecticide on the outside of buildings and sealing cracks and crevices around doors and windows. However, there is little evidence that caulking alone can significantly reduce the number of insects that enter.

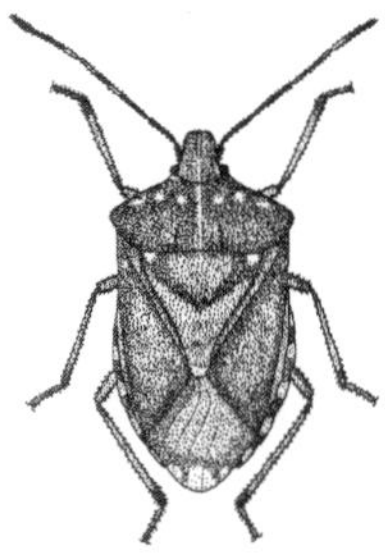

Stink bug

PHYSICAL REMOVAL

The physical removal of rodents, birds, and large animals from a location can be an effective control method. Populations of these pests are usually small, and removal can provide long-term relief. However, trapping is not always successful for controlling insect pests. Populations are usually large, and there are natural reservoirs that replenish what is removed. Trapping in large insect infestations would provide only limited reduction. However, when populations are small, and trapping is extensive, population reduction can be achieved.

Trapping or any other removal method alone, without other control methods, rarely provides long-term reduction in insect and mite infestations.

Vacuuming. Vacuuming is often recommended as a control strategy for infestations of cat fleas, bed bugs, and German cockroaches. When infested harborages are known and accessible, removal seems a logical control strategy. However, removal of these pests provides only temporary reduction of the infestation. The primary reason for this is that not all harborages are accessible to vacuuming, so a portion of the infestation will remain. For example, vacuuming for household fleas can remove a large percentage of flea eggs from carpet, but only a small percentage of larvae.

- *Dust mites.* Live dust mites are found at sites in houses which have an abundance of human skin scales and more than 45% relative humidity (RH), such as beds, soft furnishings, and carpeting. Mite populations in carpets can range from 200,000 to 500,000 live mites per 10 square feet. Although vacuum cleaning is often recommended for reducing infestations, there is little evidence that even powerful machines can significantly reduce populations or decrease allergen levels.

Vacuuming can remove about 5% of the live mites present in a carpet. The majority of mites are out of the suction range of the cleaner; they cling to carpet fibers and are not removed.

Cat flea larvae feed and remain at the base of pile carpet

- *Fleas.* The pile of modern carpeting provides conditions for the accumulation of food for cat flea larvae, such as the dried blood of a host and organic debris. Carpeting is a microclimate ideally suited for the development of flea larvae. It provides a harborage that limits insecticide penetration and vacuum removal. Larvae are usually at the base of the pile, and with the aid of long setae on their body, they resist the action of the vacuum.

- *Bed bugs.* Vacuuming can be effective in eliminating small infestations of bed bugs. However, for large infestations, vacuuming alone can not provide elimination. Many bed bug harborages, such as wall voids, picture frames, and electrical devices are inaccessible to vacuum cleaner tools. Adults and nymphs are difficult to remove from fabric and other rough surfaces.

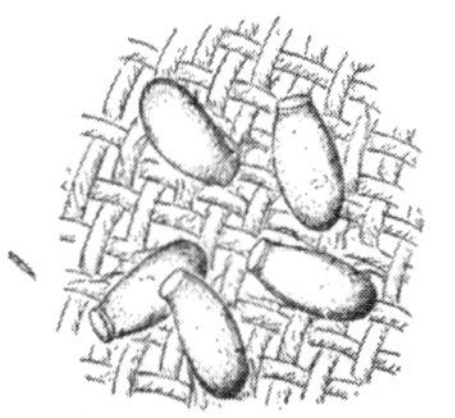

Bed bug eggs

Bed bug eggs are covered with an organic substance when they are laid. This material adheres to surfaces, which makes the eggs difficult to remove.

The immediate value of vacuuming areas infested with bed bugs is removal of dirt and debris, which can make it easier to inspect and detect bed bug activity.

Vacuuming can remove about 95% of adult fleas and 90% of the flea eggs that are in carpet pile, but extensive vacuuming removes only about 27% of flea larvae.

Travel through the vacuum cleaner brushes and extension hose to the collection bag is lethal to adult fleas and larvae removed from carpeting.

- *Cockroaches.* Commercial vacuum cleaners may be used for removing German cockroaches from harborages. The use of a flushing agent and vacuuming can be effective in reducing large infestations. Flushing and vacuuming six to eight weeks after insecticide application can provide about 8% additional reduction in infestations. A significant benefit is the removal of females carrying egg cases.

Food Attractant Traps can be helpful in the control of:

- *Yellowjackets.* Colonies of yellowjackets have a division of labor in which a large number of individuals are responsible for gathering food. Food lures and commercial attractants can be used in live-capture traps to remove foragers from colonies that are close to public areas. Removing a large number of foragers can weaken colonies and reduce the threat of stings. A large number of traps can be placed at the periphery of an area or outdoor building to reduce the number of foraging yellowjackets by intercepting them before they encounter people.

  The traps hold a small amount of bait or food attractant in a container that allows access through 3/8- to 1-inch holes. Effective trap colors are translucent white, green, and yellow, or a combination of these colors. Bait materials include fish, ground beef, sugar solutions, and commercial yellowjacket attractant (n-heptyl butyrate).

Yellowjacket

- *Rodents.* Rats and mice visit familiar sites where food is available and are attracted to food odors. Rodenticide baits rely on this behavior by combining desirable food with a toxic ingredient. The non-toxic rodent monitoring or tracking blocks can be used in snap-trap stations. Rats and mice are attracted to these stations by the highly palatable, non-toxic baits. Once they enter the station, the lethal snap trap is triggered. Other non-toxic baits or liquid attractants are available for use with snap traps and in bait stations.

Sticky Traps. Modern sticky traps are available in a range of sizes and configurations. The design and use of food and pheromone attractants in modern sticky traps can make them an effective control tool. When infestations are small or confined to a limited area, an extensive trapping program can be effective in reducing and eliminating pests.

Professional sticky trap

Indian meal moth

Mediterranean flour moth

- *Cockroaches.* A professional sticky trap designed for cockroaches can function as a control tool. These traps, with 30-degree angled entry ramps and a food attractant, can help attract and remove a large number of individuals. Traps are most effective when females with egg cases and large nymphs are removed from the population. Females with egg cases are responsible for increasing the infestation, thus removing large numbers of them can result in a decline of an infestation.

- *Stored-food pests.* Trapping can be used as a control strategy for moth and beetle pests. The objective is to remove a sufficient number of adults or to disrupt mating to achieve population reduction.

Mating Disruption Traps. These traps permeate the environment with a sex pheromone, which results in males being unable to locate and mate with females. They can not follow an isolated "pheromone trail" to a female, because the environment is saturated with pheromone. Unmated females do not reproduce, and the population begins to decline.

- *Mass trapping.* This "trap-out" strategy places aggregation pheromone traps throughout the habitat to remove a large number of males or females. The objective is to quickly reduce the population. This strategy may only be practical in small and closed facilities and can be expected to provide only short-term relief. Mass trapping is feasible when pest populations are small.

At least 90% of the males in a meal moth population must be removed to achieve significant reduction; however, many of the male moths attracted to a pheromone trap are not caught.

- *Lure-and-kill.* This is a modification of the mass-trapping technique and has been described as attract-and-kill, or attraction-annihilation. In this strategy the flour moth or meal moth is lured by a pheromone but is not caught in a trap. Instead it is subjected to a killing or sterilizing agent that eliminates it from the population.

UV Light Traps. Insects are generally attracted to the brightest light they see. The brightest lights for insects are in the ultraviolet (UV) portion of the spectrum. UV light is out of the visible range for humans and has a wavelength of 100 to 400 nanometers (nm). UV light traps are used primarily to control house flies around food. Glue boards positioned behind or below the UV bulbs capture insects landing near the bulbs, or electrified metal grids in front of the bulbs kill insects flying to the light. Both methods are effective in removing pests from enclosed spaces. However, neither can be used alone in a control program.

- *UV Bulbs.* The typical UV light bulb has a light output concentration of 340 to 360 nm. House flies are attracted to light at 365 nm; fruit flies are attracted to 340 to 360 nm. Standard UV bulbs have a peak of green light (near 550 nm); this light is attractive to some stored-food moths.

- *Shatter-proof.* In food- and drug-manufacturing sites, shatter-proof bulbs are required. This is accomplished by sheathing standard bulbs in clear plastic. Depending on the polymer plastic material used, there can be a slight reduction of the total UV output of the bulb. Because of the amount of heat at the ends of the bulbs, plastic sheaths may darken and become brittle.

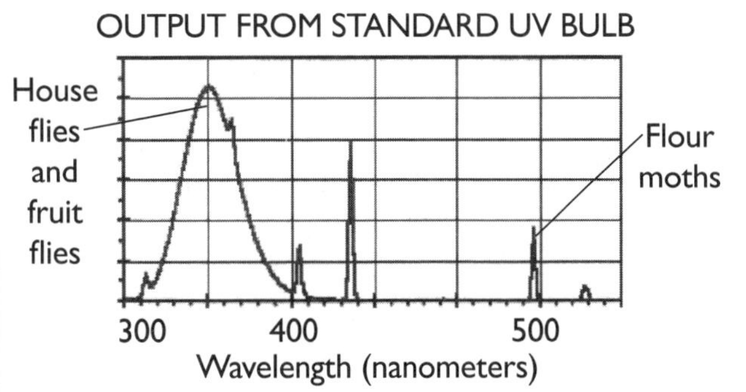

- *Electrocution traps.* UV light traps designed to electrocute flies are used in kitchen, food preparation, and dining areas. These traps are often placed outside doors and in entryways to intercept flies as they enter. When used indoors, they are placed so that dead insects from the traps do not fall on food, food contact surfaces, equipment, or utensils. There are electrocution traps that mildly stun the insect so that it falls intact onto a glue board, and those that kill, which often results in fragmenting the insect. The house fly is susceptible to fragmentation because of its thin cuticle.

- *Bulb replacement.* The ultraviolet output of UV light bulbs decreases approximately 30% in the first 30 days of use. Within six to eight months, bulbs have lost more than 50% of their UV light output and are ineffective in attracting insects. Since UV light is invisible to the human eye, there is no change in the outward appearance of the bulb. The loss of UV output can only be measured by a decrease in capture or the use of a UV-A light detection meter.

Standard practice in food plants is to have no less than 10 feet between electrocution-type light traps and food materials.

Bulbs in light traps used for house fly control should be replaced at the beginning of the fly season, so that peak UV light output coincides with the peak in fly activity.

- *Position.* Traps with horizontally directed light beams will capture nearly 1.5 times as many flies as traps with downward-directed beams, and three times more than traps with upward directed beams.

- *Attraction.* Most insects have two peaks of maximum light attraction: one in the ultraviolet wavelength of about 350 nm, and a second in the blue-green of about 500 nm.

- *House flies.* For house flies and blow flies, the dominant peak is in the ultraviolet wavelength of 350 to 365 nm. Their age, gender, and food intake can influence their response to UV light. Young flies are more attracted than those that are five days or older; and unfed female flies are more attracted to UV light than fed females.

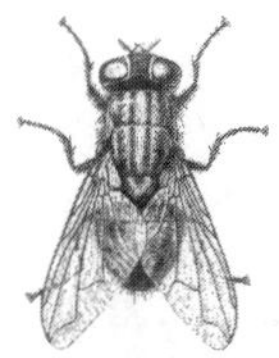

House fly

House flies are attracted to green light, but at temperatures below 75°F. Some stored-food moths are also attracted to green light.

It can take nearly 3 hours to achieve a 75% house-fly capture with an UV light trap, and more than 5 hours for a 90% capture. Regardless of the fly species, or the number of UV light traps used, 3% to 10% of the flies in a room may remain uncaught.

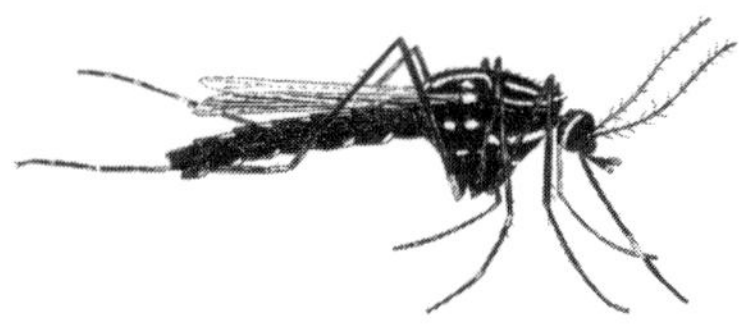

Mosquito

Carbon Dioxide Traps. The attractant in mosquito traps for home use is usually based on emissions of a small amount of CO_2, which lures female mosquitoes. Some traps add lights and heat, and the chemical octenol, which is exhaled by mammals.

The efficacy of CO_2 traps is limited to mosquitoes that typically breed around structures. These traps can reduce mosquito biting in a small area for a limited time if there is little wind to disrupt the attractant cloud of chemicals. They work best when mosquito numbers are relatively low, and the traps are placed 30 to 45 feet from the area to be protected. Mosquitoes can detect the motion of animals within approximately 30 feet. If traps are placed too close to people, some of the mosquitoes drawn to the trap may be diverted to people instead.

Sanitation may decrease harborages and force pests to increase their foraging activity. This increased pest activity may then increase their contact with chemical residues.

SANITATION

Control programs for cockroaches, bed bugs, and other household pests usually include sanitation. Sanitation is broadly defined as reduction of available food, water, and harborage in the infested site. It may be difficult to eliminate these from households, but any decrease will put stress on pests and limit their survival.

- *Cockroaches.* Sanitation can be used in cockroach management programs, but it, alone, can not provide a suitable level of control. While abundant food and water favors large infestations of German cockroaches, there is little relationship between the sanitation level and the number of cockroaches in infested locations.

Sanitation as a control strategy may not significantly decrease cockroach numbers. It does not reduce food, water, and harborage sufficiently to affect the population.

- *Bed bugs.* Sanitation can contribute to effective insecticide application for bed bug control because it reduces the treatment sites. Excess clothing, furniture, wall hangings, and clutter in bedrooms provide harborage for adults and nymphs and egg-laying sites for females. Bed bug elimination depends on treating of all harborages with a residual insecticide or using heat, steam, or other method. Removing, cleaning out, or washing excess material will likely contribute to long-term control.

Clean bed bugs from items such as clothing, area rugs, stuffed animals, and backpacks by placing them in a clothes dryer for five minutes at a low (140°F), medium (150°F), or high (180°F) setting.

Bedding and clothing can be cleared of bed bug adults, nymphs, and eggs by washing in a household washing machine using hot water and the standard wash cycle, followed by drying at low to high heat in a dryer.

- *Rodents.* The attraction of rodents to bait is subject to environmental conditions. Sanitation in the form of removing alternate food can encourage rodents to forage for new sources and to feed on baits. Sanitation measures may precede the placement of bait stations to direct rats to the food bait when their existing food is removed.

HEAT

Insects tolerate only a narrow range of temperature. At temperatures of 100°F to 120°F, an insect will die in 24 hours; above 120°F it will die in minutes. Long exposure to high temperatures is usually lethal, because it results in water loss or damage to vital organs.

Heat treatment techniques depend on the effective distribution of the heat, mobility of the target pest, and sensitivity of household components to high temperatures. In a heat treatment, the air within the structure is gradually heated to 120°F and 140°F.

- *Termites.* Thermal treatment is used for drywood termite control and is usually conducted as a spot treatment to the site of the infestation. The extent and location of the colony and galleries is first determined, then heat is applied to these areas. The heater usually remains outside the building with fans directing the heated air through ducts. Fans inside the building distribute the hot air to raise and maintain the ambient temperature in the structure to 150°F to 170°F.

- *Bed bugs.* Hot air is lethal to bed bug eggs exposed to 113°F for 1 hour, and to adults exposed to 118°F for 90 minutes or 122°F for 1 minute. Professional pest control methods typically use 120°F. Heating is usually progressive to prevent surface temperatures from rising faster than 15°F per hour. Some household wood laminates may be damaged if heated quickly.

The efficacy of heat treatments for bed bugs can be limited because adults and nymphs can move deeper into harborages to avoid the heated air. There may be cracks and crevices in furniture, bed frames, and behind baseboards that are not reached with the lethal air temperatures.

A common misconception is that bed bugs can be killed by enclosing infested materials in black plastic bags and placing them in the sun. This is ineffective, especially for large items such as mattresses and box springs, because the temperature will not be raised high enough to kill the bed bugs.

- *Lice and fleas.* Hot air is lethal to the eggs of lice exposed to 122°F for 1/2 hour, and to adults exposed to 115°F for 1 hour. Exposure to dry air (30% RH) for at least 1 hour at 68°F is lethal for flea larvae. In moist air (90% RH), flea larvae are killed at temperatures between 97°F and 102°F after 1 hour exposure.

- *Stored food pests.* Stored-product beetles are killed when exposed to 104°F for 24 hours, 113°F for 12 hours, 122°F for 5 minutes, 131°F for 1 minute, or 140°F for 30 seconds. Temperatures between 97°F and 102°F are lethal for mealworm larvae. Larvae of the cigarette beetle are tolerant of heat, while the larvae of rice weevils and red flour beetles are moderately tolerant. Acclimation to heat can result in survival at high temperatures.

- *Cockroaches.* Heat can be used to eliminate cockroaches from commercial structures. Exposure for 1 hour in moist air (90% RH) of 109°F is lethal for adult American, German, and Oriental cockroach males. The process for using heat involves caulking or sealing of holes in walls and exhaust vents and turning off all electrical equipment.

- *Dust mites.* Dust mites are found in carpets, upholstered furniture, and mattresses. They prefer to remain near the surface of a mattress, usually within the outer inch. Human skin scales, that are the food of dust mites, generally do not penetrate deeper than 1/2 inch into a mattress. House dust mites are killed when the relative humidity drops below 45%.

Dust mite populations in beds can be reduced by as much as 84% by heating the mattress with an electric blanket.

STEAM

The most practical application of extreme heat is to combine it with water in the form of steam. The advantage in adding steam is that it will usually kill all life stages of insects, including the egg stage. Steam as a control strategy has been used as a part of bed bug elimination and dust mite management programs. The pest habitats targeted are the mattress and carpeting.

- *Bed bugs.* Commercial steam application machines that produce steam at a low vapor flow and high temperature are the most effective. The steam emission tip is held about an inch from the surface. If the tip is not close to the surface, the steam may not be hot enough to kill the pest. The lethal temperature for steam treatment is 122°F. Steam is usually coupled with insecticide application for successful elimination of bed bugs.

- *Dust mites.* Commercial steam cleaning of household carpeting can significantly reduce the number of live mites and mite allergens in the carpet. After steam treatment, there can be a mean reduction of 87% of the mite-allergen concentration in carpeting.

Cold temperatures have lethal effects on bed bugs. Exposure to 5°F for 2 hours is lethal to bed bug adults. If acclimated, bed bugs can survive at or below 32°F for several days.

COLD

Low temperatures have been used to control pests in paper, wood, hide, and leather. The commercial use of liquid nitrogen to control drywood termites is approved in some regions. Controlling insects infesting framing timber requires introducing liquid nitrogen into the wall void adjacent to the infested wood. The time required for wood to attain the lethal temperature depends on the application point and the nitrogen flow rate.

ELECTRICITY

High Voltage Electricity can be used to control drywood, but not subterranean, termites, powderpost beetles, and carpenter ants. This method can be used alone or in conjunction with other methods, such as localized chemical treatments.

A hand-held device is available that uses a 90,000-volt electric current. The mode of action is to penetrate the wood and electrocute termites in galleries below the surface. The operation of this gun involves high frequency and high voltage, but low current, to overcome wood's natural resistance to the flow of electricity. It is the current that penetrates the wood and flows into the termite galleries and nest sites. Most insects are killed instantly, but there can be delayed mortality for those not directly contacted.

Electronic Devices. Some consumer products claim an unsubstantiated ability of electrical currents to produce sonic or ultrasonic sounds, vibrations, or electromagnetic fields to control household pests, including insects, spiders, mice, and rats.

There is no evidence or supporting research to show that electronic devices provide control of any pest, indoors or outdoors.

AIR CURRENTS

Strong air currents can prevent flying insects from entering buildings, rooms, or other spaces. Air "curtains" are created by directing a narrow band of air from overhead to a horizontal floor surface below. The air stream must maintain its strength and direction until it reaches the floor. The efficacy is based on the principle that a flying insect can not pass through the stream of downward directed air. In general, air curtains are used to keep flies and other small insects out of food storage, preparation, and serving areas.

Large insects, such as wasps and some beetles, may be able to penetrate air curtains designed for flies. Air curtains must exceed the normal flying speed of the pest: fruit fly, 6.7 mph; house fly, 9 mph; and honey bee, 15.6 mph. German cockroaches are repelled by air velocities greater than 4 mph, and American and smokybrown cockroaches by 9 mph.

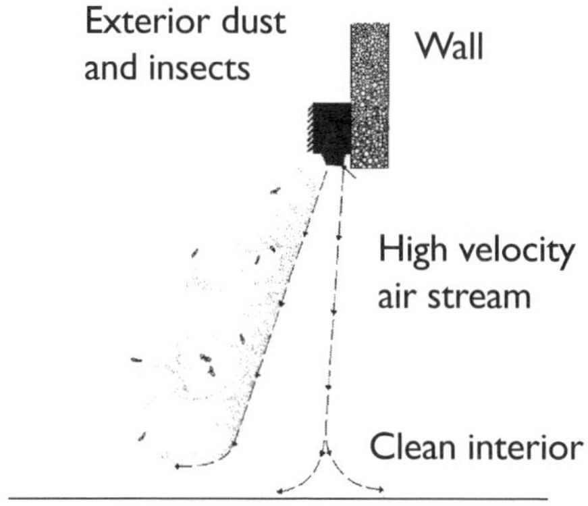

OIL AND CHEMICAL VAPORS

The toxicity and repellency of red-cedar vapor to clothes moths is a popular belief. However, investigations have shown little to no repellent value of cedar-oil vapor or cedar chests in protecting clothes. Vapor can control the newly hatched and half-grown larvae of clothes moths, but will not control adults or eggs. Cedar chests can inhibit the hatching of clothes moth and black carpet beetle eggs, but have no effect on eggs deposited outside the chest and then introduced. The efficacy of white cedar dissipates after about 16 months, and the efficacy of red cedar sharply declines after about 24 months.

Clothes moth

Exposure of clothes moth adults to the cedar in chests results in some mortality, but adults will mate and lay viable eggs.

Cedar flake board is somewhat repellent to German cockroaches, but not repellent to either American cockroaches or smokybrown cockroaches.

The repellency and toxicity of the vapor of paradichlorobenzene (PDB) mothballs has long been used to protect against clothes moths. At one time mothballs consisted primarily of naphthalene, but due to its flammability, modern mothballs use PDB. This volatile chemical kills adult moths and larvae as its solid form vaporizes. To be effective, mothballs need to be placed with clothing in a sealed container. In addition to repelling insects, mothballs are used to repel snakes, mice, or other pests.

PDB vapors are not repellent to clothes moth adults and larvae, but the vapors are toxic to eggs, larvae, and pupae of clothes moth when confined in an air-tight container.

Paradichlorobenzene and naphthalene will damage natural rubber products (but not synthetic rubber), and exposure to solids or vapors should be avoided.

CONTROLLED ATMOSPHERE TREATMENTS

Controlled atmosphere treatments are produced artificially and are maintained by enveloping an object or structure with a gas such as carbon dioxide or nitrogen. The atmospheres most commonly used are: 60% CO_2 coupled with limiting the oxygen concentration in the air space to 10% to 20%, and hypoxic nitrogen atmosphere, which has less than 0.5% oxygen with the remainder being nitrogen.

Treatment with 100% CO_2 can provide 100% mortality of all stages of bed bugs within 3 hours. An air-mixture of 60% CO_2 produces the same result in 24 hours.

CHAPTER 4

CHEMICAL INSECTICIDES

Most insecticides interfere with the functioning of the nervous system. Some disrupt energy production in the cell, while others influence the hormones that regulate development and molting, or alter the formation of the cuticle. Insecticides can enter the insect body following direct contact with a liquid or dust residue, or exposure to lethal gas or vapor. Some are absorbed through the digestive system, others through the mouthparts after grooming of tarsi and antennae.

The insect's body covering, nervous system, and metamorphosis (the process of development from immature to adult) are the biological systems targeted by insecticides. A review of these systems is necessary before considering the classes of insecticides. Understanding how silica aerosol dust kills cockroaches or bed bugs requires some background on the wax layers covering insects. Knowledge of insect development helps explain why insect growth regulators do not provide immediate kill, but rather, work behind the scenes to affect insect development. The insect nervous system is a broad target for modern insecticides. A basic explanation of the key components of this system will help in understanding mode of action of the new generation of insecticides.

INSECT INTEGUMENT

The integument, or body covering, of the insect is composed of a series of thin layers of waterproofing and sealing materials. The primary role of the integument in arthropods is to form a supporting and protective shell or exoskeleton. It also prevents water and other liquids from entering or leaving the body. Water loss prevention is important to insect survival because of its small size. Small animals have large surface areas per volume, and this increases the potential for lethal water loss.

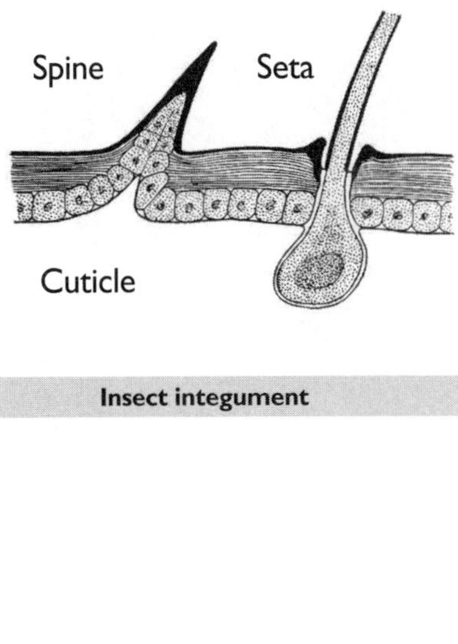

Insect integument

The major parts of the integument are:

- *The cuticle.* The outer portion of the integument is called the cuticle, and it is divided into several distinct layers. The outermost is the cement layer; it is hardened similar to shellac and protects the soft wax layers beneath it. The cement layer is vulnerable to damage; it can be scratched by an abrasive insecticide dust. The layer of wax beneath is thin and controls the water loss and chemical permeability of the cuticle.

Removal of any part of the wax layer can result in lethal water loss; a 30% loss of body water is lethal to all arthropods.

- *The chitin.* This is the major constituent of the cuticle of insects. During molting, when the insect sheds its old exoskeleton, chitin in the old exoskeleton is dissolved by enzymes so that it can be reused. Other enzymes help to form the new cuticle. Chemical compounds that inhibit these enzymes can be effective as pest control agents.

INSECT NERVOUS SYSTEM

The insect central nervous system consists of a brain in the head and a chain of small nerve centers (ganglia) along the ventral side of the body. Nerves from ganglia extend to sense organs and muscles in each body segment. A group of small ganglia control the digestive system.

The basic elements of the nervous system are nerve cells or neurons. Small branches, known as dendrites, extend from the nerve cell and receive messages or impulses from other nerve cells. A long axon conducts these impulses between nerve cells.

The ends of the axons from two cells do not contact each other; instead there is a gap or synapse between the end of one axon and the next.

The activity site for most modern insecticides is at the synapse, where the chemicals carry the nerve impulse from one axon to the next.

Transmission at the Synapse. Transfer of an impulse from one axon to the next involves crossing the synapse. Bridging the synapse is done by the chemical transmitter, acetylcholine (ACh). The end of the axon contains tiny vesicles, each containing ACh. When an impulse moves to the pre-synaptic end of the axon, it triggers the vesicles to release ACh into the synapse. ACh moves across the synapse and attaches to receptors on the post-synaptic side. This allows the impulse to flow into the axon on the adjacent nerve cell. The action of ACh is brief because acetylcholine esterase (AChE) rapidly inactivates it.

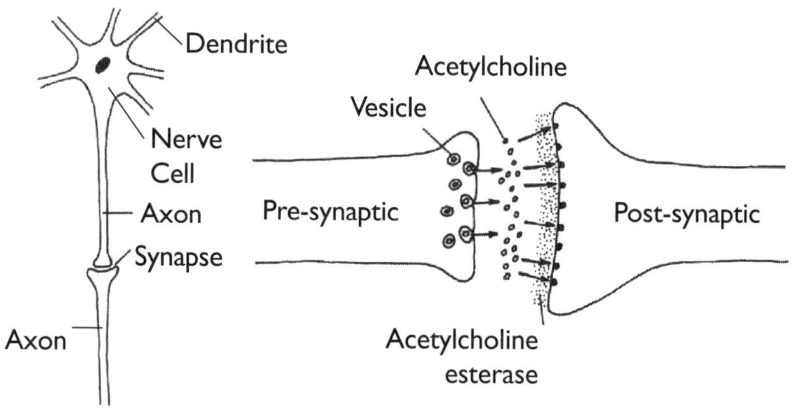

INSECT METAMORPHOSIS

The insect's change in form during development, from egg to adult, is called metamorphosis. There are two basic types:

- *Gradual or incomplete metamorphosis* – the development is a succession of nymph stages, in which each is a larger version of the previous stage.

- *Complete metamorphosis* – development includes a dramatic change from several larval stages to a pupa, and then to the adult.

  The changes during metamorphosis are controlled by ecdysone and juvenile hormone. The process of molting, growth, and development is triggered by ecdysone. Juvenile hormone (JH) regulates the change from larva or nymph to the next stage. The amount of JH naturally decreases during the last immature stage to allow the development of larva to pupa, or nymph to adult.

The artificial presence of JH, such as from an insect growth regulator insecticide, induces the production of an additional and lethal larval or nymph stage.

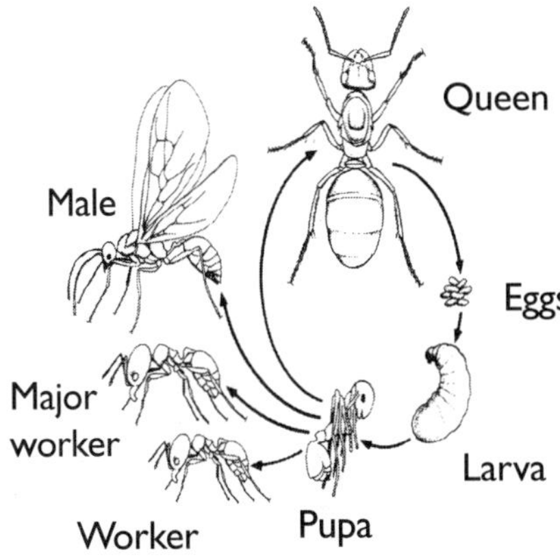

Ant metamorphosis

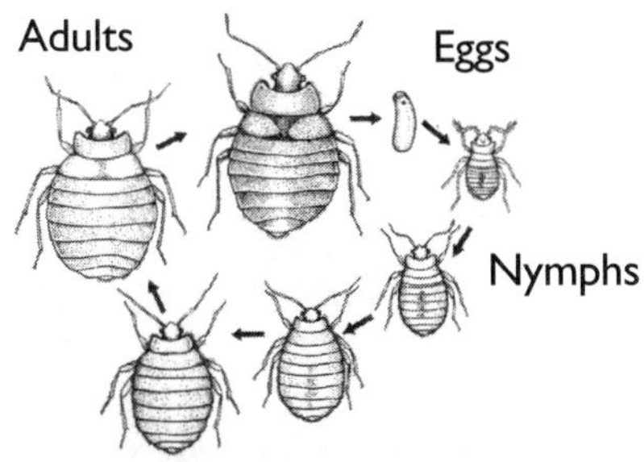

Bed bug metamorphosis

INSECTICIDES

Household and structural insecticides target sites in the insect nervous system. The organophosphates and carbamates attack the nervous system, but they are slow acting in comparison to the pyrethroids. The pyrethroid insecticides are low-odor and provide a flushing action and quick kill. Cockroach baits are typically slow-acting and kill within three to four days, when the insect is in its harborage.

The neonicotinoids and oxydiazines target the nervous system but they have a different approach. While the pyrethroids open the sodium channels in the nerve fibers, the oxydiazines block these channels. The organophosphate insecticides allowed acetycholine (ACh) to accumulate at the nerve synapse, the neonicotinoids add ACh to the nerve synapse. Although the mechanisms seem similar, the new compounds are effective at much lower concentrations and with no vertebrate toxicity.

Pesticides are classified by the Environmental Protection Agency (EPA) according to established risk criteria.

- *Reduced Risk status.* The criteria established for pesticides to have Reduced Risk status include low impact on human health, low toxicity to non-target organisms, low potential for groundwater contamination, and low use rates. Many pesticides derived from natural materials, such as plants, bacteria, and certain minerals pose a low risk to human health. The low mammalian toxicity of some of the modern insecticides, such as some of the neonicotinoids and diamides also enables Reduced Risk status.

- *Minimum Risk status.* The EPA considers it unnecessary to regulate a group of chemical pesticides that are based on ingredients considered to have minimum risk to humans and the environment. To qualify for Minimum Risk status each active ingredient and inert ingredient in the pesticide must be listed by the EPA. The list is updated on a continuous basis on its Pesticides website. (Current versions are on the EPA website, www.epa.gov.)

Avermectins. These are natural products of fermentation of the soil microorganism, *Streptomyces avermitilis*. Avermectins cause extreme excitation at the junction of nerves in the muscles of insects. They block the activity of muscles in insects. Abamectin insecticides are a mixture of avermectin B1a and B1b.

Bacteria. Toxins produced by the bacteria *Bacillus thuringiensis* cause cells to rupture and disintegrate. When ingested by insect larvae the bacteria release toxins into the digestive system. The toxins destroy the lining of the gut and feeding stops. The acid conditions (4.8 to 6.3 pH) of the cockroach digestive system may be the reason for the lack of activity of *B. thuringiensis* on these insects.

Botanicals. This group of insecticides is based on plant oils. Of the active ingredients that are classified by the EPA as having Minimum Risk exemption, most are botanically derived essential oils. The list of Minimum Risk active ingredients includes: castor oil, cedar oil, cinnamon oil, citric acid, citronella oil, cloves and clove oil, corn gluten meal, corn oil, cottonseed oil, eugenol, garlic oil, geranium oil, lemongrass oil, linseed oil,

The intense odors of some essential oils are often correlated with the level of biological activity, but this is not always the case.

malic acid, mint oil, peppermint oil, rosemary oil, sesame, sodium chloride, sodium lauryl sulfate, soybean oil, and thyme oil.

- *Limonene (d-Limonene) is extracted from citrus fruit peels.* It stimulates nerve cells in the nervous system to produce repetitive discharges and eventually causes paralysis. Limonene does not seem to affect the central nervous system of insects.

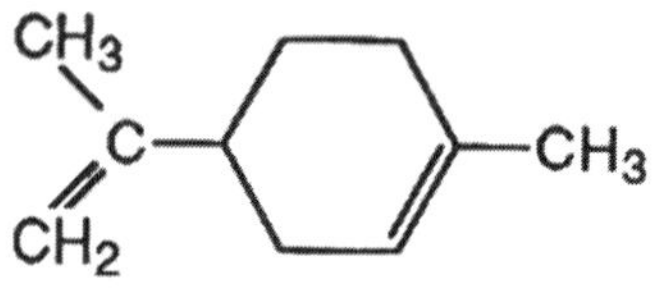

d-Limonene

D-Limonene is not compatible with Buna-N or natural rubber, neoprene or nitrile gaskets, or polyethylene or polypropylene spray tanks. Thus, stainless steel tanks must be used.

- *Geraniol is an alcohol extracted from geranium oil or citronella oil.* This plant material is repellent to mosquitoes, house flies, cockroaches, ants, fleas, and ticks; however, it can attract honey bees.

- *Thyme oil (thymol) is the active ingredient in some ready-to-use insecticides and disinfectants.* It is highly lipid soluble and may penetrate the insect cuticle when used as a contact spray.

Boric acid is a stomach poison, but it will penetrate the cuticle. When only exposed to the cuticle, mortality of the target pest may be delayed.

Borates, Boric Acid. In high concentrations, borate compounds are toxic to cells of plants and animals. However, these are slow-acting insecticides; the lethal effect on insects is achieved through contact or ingestion. Boron interferes with the process of breaking down food, such as glucose, into energy. The insecticidal effect of boric acid is achieved through a high concentration of boron in the body. Once ingested, boron moves into the circulating blood; for cockroaches this requires about 24 hours.

Diamides. These insecticides bind to ryanodine receptors in muscle cells, which results in massive muscle contractions and death. For example, exposure to chloroanthraniliprole causes a disruption of muscle contraction, which results in the insect losing control of its muscle systems. The distinct difference between the ryanodine receptors of insects and mammals provides for very low mammalian toxicity. This low toxicity qualifies these insecticides for the EPA Reduced Risk status, which is the lowest toxicity rating.

Worker termites exposed to chloroanthraniliprole may continue to move for a short time, but they are unable to tunnel or feed. Exposure may be from contact with treated soil or residue transferred from nestmates.

Oxadiazines. Oxadiazines attack the nervous system as sodium channel blockers. Indoxacarb undergoes metabolic "activation" once it enters the insect body. This is a unique mode of action, since metabolic actions normally degrade or inactivate insecticides in the body. These insecticides are non-repellent and are generally used at low concentrations. Indoxacarb, dinotefuran, and other oxadiazines have Reduced Risk status.

Cyclodienes. This group includes the insecticides aldrin, chlordane, dieldrin, endrin, and lindane. The cyclodienes function by blocking the GABA-controlled chloride channel in the nerve. Although resistance to chlordane developed early in German cockroaches and other pests, the chloride channel remains a target site for some modern insecticides. For example, fipronil successfully blocks the GABA-controlled chloride channel.

Desiccant and Abrasive Dusts. These dusts abrade the surface of the cuticle or adsorb the wax layer. This results in the loss of a lethal amount of body water. The insecticidal action of these materials is slow. The commonly used products are diatomaceous earth, silica aerogel, and finely ground limestone.

- *Silica aerogel is made from a geological deposit composed of fossilized skeletons of marine and freshwater animals.* These skeletons are made of silica, and when crushed, they break up into fine, talc-like fragments. Lightweight silica with low bulk density and high porosity are called aerogels. Their mode of action is the ability to adsorb waxes from insect cuticle, which permits excess water loss from the body. To increase efficacy of the extremely lightweight silica aerogels, they are formulated with petroleum oil, pyrethrins, and a synergist.

- *Diatomaceous earth is principally silica, and it acts as an abrasive.* It has only limited ability to adsorb the wax from insect cuticle, and insects vary widely in their response to this desiccant. Some species have a protective cement layer in the cuticle, which provides some protection against dust desiccants.

Pyrethrins. These chemicals are extracted from flowers of chrysanthemum plants. Pyrethrin-based insecticides interfere with the transmission of impulses along the neuron. This results in repeated excitation of the nerve and causes paralysis in the insect. Pyrethrins produce flushing, rapid knockdown, and, at high concentrations, killing action. Some insects recover from exposure of low doses, but this can be prevented by combining piperonyl butoxide with the pyrethrin. Pyrethrins have low toxicity to mammals, and have short residual activity because they degrade in sunlight.

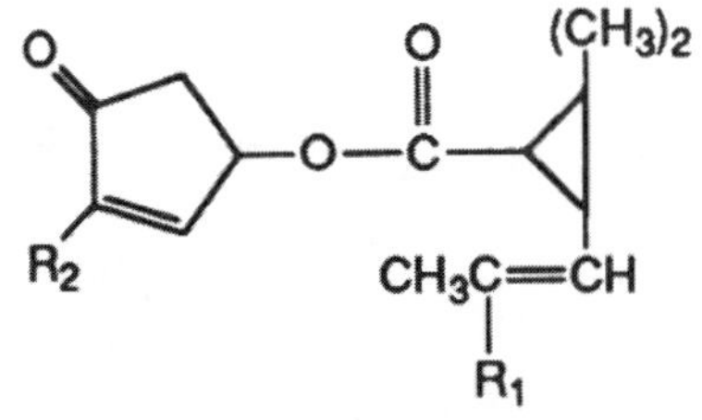

Pyrethrin

Carbamates. These are cholinesterase inhibitors, except for fenoxycarb which acts as an insect growth regulator. Carbamate insecticides have low mammalian toxicity and a broad spectrum of insect control. Bendiocarb is effective against a range of insects, especially ants, wasps, fleas, ticks, and mosquitoes. Propoxur is used in consumer and professional aerosols. Methomyl is an active ingredient in fly bait products.

Organophosphates. Organophosphates interfere with the transmission of nerve impulses at the synapse. They block the action of acetylcholine esterase (AChE) in clearing away acetylcholine (ACh). Accumulation of ACh at the synapse is toxic to insects. Organophosphates for household pest control are: acephate, dichlorvos (DDVP), fenitrothion, and propetamphos. Malathion is one of the oldest and most widely used of the organophosphates. Because of its very low mammalian toxicity, it can be broadly applied for public health pests, especially as an aerial spray or thermal fog for mosquitoes.

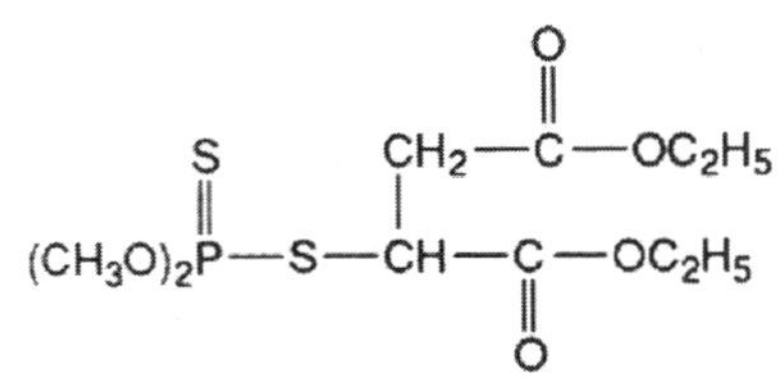

Malathion

Pyrethroids. These are synthetic analogues of natural pyrethrins. They are effective at low doses, and are generally characterized as having low odor and quick knockdown and kill. In general, the pyrethroids have low mammalian toxicity. Pyrethroids are biodegradable, so there is no buildup in the food chain or the soil.

Although the first pyrethroid (allethrin) was made in the 1940s, the first generation of general-use pyrethroids appeared in the 1960s, represented by

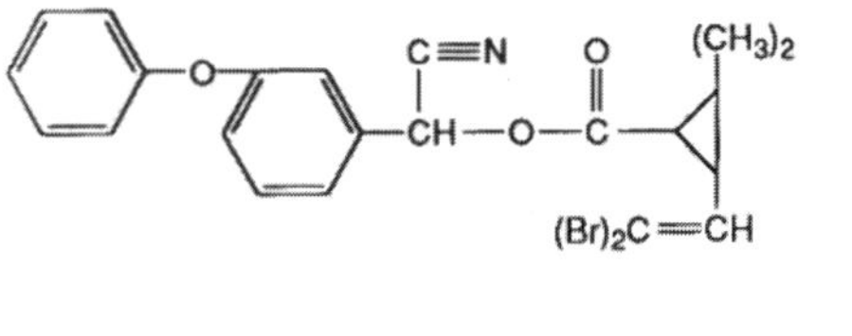

Deltamethrin

tetramethrin and bioresmethrin. The second generation of residual and photostable compounds came in the 1970s, represented by cypermethrin, deltamethrin, and fenvalerate. The most commonly used include bifenthrin, cyfluthrin, cypermethrin, deltamethrin, esfenvalerate, fluvalinate, and lambda-cyhalothrin. In addition to knockdown activity, the pyrethroids metofluthrin, transfluthrin, and d-trans-allethrin, inhibit mosquitoes from biting, which occurs well before knockdown.

Fluorosulfonates. These unique insecticides are stomach poisons; they have delayed-action toxicity. The mode of action is depression of the rate of oxygen consumption in individual cells, which denies cells of needed energy. Sulfluramid-based insecticides are effective against termites, cockroaches, ants, and other household and structural insect pests.

Phenyl Pyrazoles. These insecticides attack the central nervous system of insects. In general, they act by hyper-exciting the nerves and muscles in the body, resulting in the insect's death. Fipronil and other compounds in this class are effective at low concentrations and control a broad range of insects.

Pyrroles. Pyrroles are derived from microbial compounds. The pyrrole insecticide, chlorfenapyr, does not become toxic until it enters the body and undergoes a change in structure. Activation is through metabolism in the insect body. These insecticides target the mitochondria in the cells, and exposure produces a gradual decline in movement and feeding, then eventually, death.

Amidinohydrazone. Amidinohydrazone and flurosulfonate insecticides, such as hydramethylnon and sulfluramid, must be ingested by insects. These insecticides are usually incorporated into toxic baits. The effect is a decrease in respiration and oxygen consumption of the body cells, which deprives the cells of energy. After ingestion, the insect stops feeding and, after 48 hours, stops all activity.

Metabolism of these insecticides in the insect body is limited, and small amounts can occur in the feces. Lethal amounts are transferred to others through their grooming, trophallaxis, and feeding on cast skins of insects.

Neonicotinoids. These are non-repellent insecticides that act on the insect central nervous system: they mimic acetylcholine (ACh) and affect the nerve synapse. Neonicotinoid insecticides block the ACh receptor sites (nicotinergic receptors) on the receiving axon. This prevents transmission of an impulse, resulting in damage to the nervous system and death of the insect. Many of these insecticides have been granted Reduced Risk status by the EPA.

The first generation of neonicotinoids, the chloro-nicotinyl group, is represented by imidacloprid and acetamiprid; second generation compounds, the thia-nicotinyl group, include thiamethoxam; and the third generation, the fura-nicotinyl group, is represented by dinotefuran.

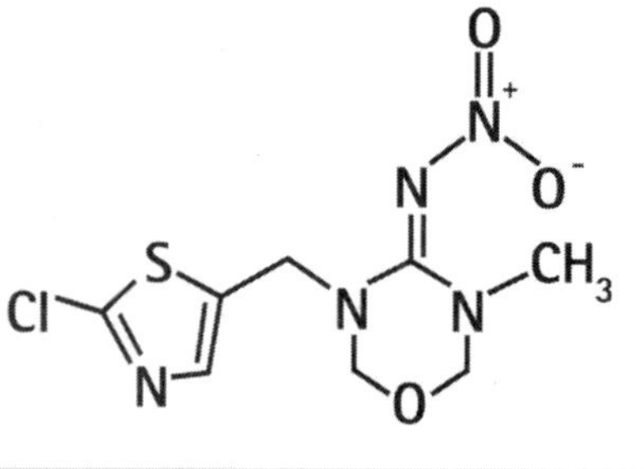

Thianicotinyl

Insect Growth Regulators. Insect growth regulators (IGRs) are chemicals that interfere with normal insect development and reproduction. They act on insects with both complete (egg, larva, pupa, adult) and incomplete (egg, nymph, adult) metamorphosis. They have no toxicity to mammals. Their mode of action is to interfere with the actions of hormones that control insect molting and development. Exposure to an IGR does not result in death, but it does cause abnormalities that prevent survival, mating, or reproduction by the adult. These chemicals are directed at an immature stage, but the results are not apparent until the adult stage. These are volatile compounds that rapidly degrade outdoors, but are persistent indoors.

- *Juvenoids.* The insect growth regulator (S)-hydroprene is a synthetic copy of a natural insect hormone. It prevents insects from completing development to the adult stage. For insects with complete life cycles, the disruption occurs at the pupal stage. The most common pests in this category are fleas, flies, meal moths, and grain beetles. For insects with incomplete life cycles, the disruption is in the last nymph stage. The result is adults with deformed wings, such as in cockroaches, and/or non-functional reproductive organs, e.g., bed bugs.

- *Fenoxycarb.* This insecticide displays juvenile hormone activity against German cockroaches and other insects. This insecticide kills larvae and pupae, and inhibits queen egg production in some ants. Pyriproxyfen is a juvenoid similar to fenoxycarb in its activity against German cockroaches.

- *Chitin synthesis inhibitors.* These chemicals disrupt the development of the insect cuticle. They must be ingested to be effective. The chemicals kill immature stages by disrupting molting, but have limited effect on eggs. After ingestion, formation of the new cuticle is disrupted and the nymph dies. Chitin inhibitors include diflubenzuron, lufenuron, hexaflumuron, and noviflumuron. Recently developed inhibitors, such as chlorfluazuron and flufenoxuron, are effective in the range of 1 to 10 ppm.

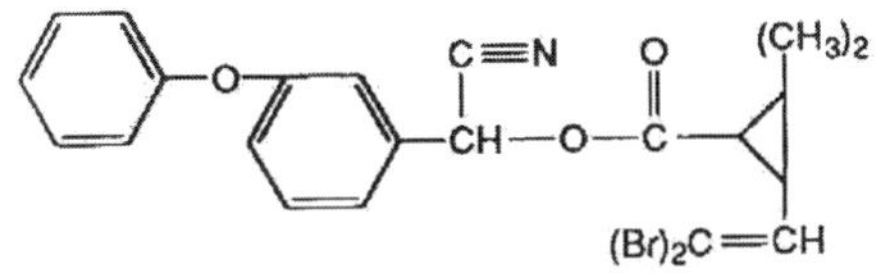

Diflubenzuron

Fumigants. Common fumigants include:

- *Methyl bromide.* This commonly used fumigant rapidly kills insects, mites, and nematodes. It penetrates substrates including soil and wood. Methyl bromide and ethylene dibromide are alkylating agents. They act at the cellular level and deactivate essential enzymes. Methyl bromide is listed as an ozone-depleting substance under the Montreal Protocol on Substances that Deplete the Ozone Layer. Developed countries are eliminating most uses of this chemical. *(See Chapter 13, Structural Fumigation.)*

- *Sulfuryl fluoride.* This fumigant is used to control a variety of household and structural pests. Use of this fumigant has been limited for the control of food pests because of the presence of fluoride as a breakdown product. The mode of action is the disruption or blocking of metabolic energy in the cells.

Mortality may be delayed for several days, as the egg shell, or chorion, of insects limits the penetration of sulfuryl fluoride, and control of eggs of beetles may require long exposure or increased concentration.

- *Hydrogen phosphine.* This fumigant is highly toxic; a concentration of 400 ppm is lethal to insects, humans, and other forms of life. Exposure of insects leads to paralysis of muscles attached to the spiracles, which prevents respiration. Without oxygen, the production of energy in the cells ceases. Exposure periods of one to five days are necessary to control most insects.

CHAPTER 5

INSECTICIDE FORMULATIONS

Most insecticide products consist of an active ingredient and one or more emulsifiers or other inert ingredients. These additional materials enhance application or improve control efficacy. The mixture of active and inert ingredients is the actual formulation; it may be in the form of a solid, liquid, or gas. Some formulations are ready to use, while others must be diluted with an organic solvent; a food material, as in the case of cockroach and ant baits; water; or air before application. The formulation influences insecticide application, persistence, transference from treated surface to target pest, and penetration into the body of the pest.

FORMULATION COMPONENTS

The performance of an insecticide is not based solely on its chemical toxicity to the pest. The active ingredient must be incorporated into a formulation that enables the user to mix and apply the proper amount. Formulations of insecticides differ by the chemistry of the active ingredient, but there are basic features such as the dilution material and chemical additives.

Active Ingredient. There may be several formulations, such as wettable powders, suspension concentrates, and dusts, for the same active ingredient. These are often determined by production costs, the anticipated short- and long-term storage environment, application equipment, and applicator safety. The chemical properties of the active ingredient also influence the formulation. In their concentrated forms, some compounds are liquids, others are solids; some are unstable in sunlight or high humidity; some are soluble in oil and/or water; and some are effective when applied as a dust, while others are not.

Mixtures. Insecticide manufacturers may combine two active ingredients into one product formulation. To accomplish this, a specific solvent system and concentration is necessary to maintain compatibility and stability of the finished material. Often, the objective of dual-action formulations is to pair a fast-acting contact insecticide with one that provides residual or long-term control. Other capabilities may be exploited, such as an active ingredient that targets indoor pests with one suited for outdoor or perimeter pest control, or an active ingredient with surface residual control with one that penetrates soil for sub-surface control.

Mixtures created by operators are likely to be chemically non-compatible because of differences in their concentrations and solvent systems. They are also likely to be in violation of state and federal regulations.

Dilution and Carrier Material. The term "carrier" is used to identify an inert ingredient that has high absorptive qualities; the dilution material (diluent) applies to ingredients with low sorptive capacities. Carriers typically have no insecticidal action, but they may have some interaction with the insecticide.

- *Water and oil.* Packaging insecticide in a concentrated form for later dilution is economical, while ready-to-use formulations are expensive. Insecticides for professional use are usually prepared as a concentrate, ranging from 5% to 80% active ingredient, and in a form suitable for dilution by the applicator, such as liquid, wettable powder, or wettable granule. The majority of products are designed for dilution in clean water; few are intended for dilution in oil-based solvents.

- *Air.* Insecticides can be delivered in small droplets as aerosols, mists, and fogs. This is done by breaking up the liquid insecticide into droplets and using air as the carrier. In aerosol applications, the droplets in the spray are small and their distribution may not be seen easily by the applicator. Aerosol sprays have droplets smaller than 50 microns, and these small droplets cannot be seen easily by the human eye.

- *Inert substrates.* Dusts may consist of the toxic ingredient alone, without dilution, as in the case of silica aerogels and boric acid. Other dust formulations include an inert carrier or dilution material, such as inorganic clay, talc, pyrophyllite, and gypsum; and organic substrates such as corn meal and cellulose. Some diluents have insecticidal properties, and some have food qualities, such as edible dust formulations for termite control.

The carriers in dust formulations are used primarily to prevent caking and clumping and to facilitate uniform application.

- *Granular formulations* may contain a small amount of insecticide, with the remainder being an organic or inorganic carrier. The inorganic carriers include clays such as attapulgite, montmorillonite, bentonite, and vermiculite. The most common organic carriers are pulverized corn cobs and walnut shells.

Dusts typically have waterproofing agents to maintain efficacy in wet locations.

Additives, Synergists, Buffering, and Foaming Agents may also be a part of pesticide formulations:

- *Additives* may enhance the stability of a formulation when it is exposed to cold and hot conditions, or facilitate extended storage and shipping. It may be an addition to the solvent system to improve the solubility of the active ingredients in emulsions and wettable powders. Additives are sometimes used to mask or neutralize the odor of active ingredients. Pyrethrum formulations usually have ultraviolet absorbent agents to extend residual efficacy. Buffering agents are additives that adjust the pH of a liquid.

- *Synergists* are chemicals that enhance the toxicity of an insecticide in a formulation. The combination of the active ingredient and synergist results in the formulation being more toxic than the active ingredient alone. Synergists usually do not increase the residual life of an insecticide. The cost of pyrethrins for insect control would be prohibitive without the advantage of synergism, as the synergists make these chemicals effective at low concentrations.

- *Surfactants* are chemicals added to water-based insecticides to facilitate their delivery as foam. They act by decreasing the surface tension of the mixing water. Decreased surface tension permits the introduction of air and formation of bubbles. The greater the concentration of foaming agent the more the surface tension decreases, which usually produces dense foam. However, each product has a specific concentration of surfactant at which the foaming effect stops. The continued addition of foaming agent can then result in clogging of nozzles and damage to the delivery system.

WATER-BASED FORMULATIONS

Most insecticides for household pests are designed to be dilutions in water. These are the easiest to mix and apply with modern application equipment. These include:

Wettable Powders, Pellets, and Granules. Water-dispersible powders or wettable powders (WP) and water-dispersible granules (WDG) are suspensions and not solutions. The insecticide is suspended in the form of micro-particles in the water, but not dissolved in water. WP formulations consist of finely ground insecticide mixed with a mineral to keep the particles separated during storage.

WP and WDG sprays are effective because of the presence of the active ingredient on the surface of the dried residue. These particles are picked up on the legs of insects walking on the surface.

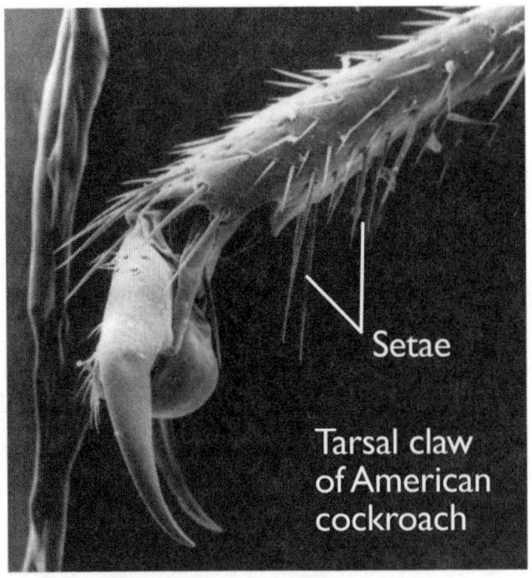

The large tarsal claws and numerous setae on the legs of cockroaches and other crawling insects dislodge insecticide particles from a treated surface.

Soluble Powders. Soluble powders (SP) are technical-grade insecticide mixed with a small amount of wetting agent. These formulations can be added directly to water to form a true solution. The powder is often very fine, and these formulations may be available in ready-to-mix dose packages, such as water-dissolvable bags or water-soluble pouches (WSP). These are easily dropped into the water in the spray tank without further mixing. The advantage is the accuracy of the resulting concentration.

Emulsifiable Concentrates. Emulsifiable concentrates (EC) are solutions that contain small amounts of emulsifying agents, which permit the suspension of minute droplets of oil in water. These formulations can be added directly to water in the spray tank. The suspension of fine droplets in the water is apparent by a white or opaque appearance of the finished dilution. Once applied to the surface, the oil and water gradually disappear by absorption and evaporation to leave the active ingredient exposed to the target pest.

Suspension Concentrates. Suspension concentrates (SC) contain finely ground solid insecticides in a liquid. The particles in these formulations range from 0.5 to 17 microns. At this size they adhere to insects walking on SC-treated surfaces. The formulations may enter the insect body directly through the cuticle or through the mouth after leg or antennal grooming. Enhanced manufacturing processes can produce formulations with a specific range of active ingredient crystals: small crystals penetrate the cuticle quickly for knockdown and kill, large crystals have residual activity.

Foam. Water-based foam is an effective means of delivering household and wood-protection insecticides. Foam will spread around objects and into crevices that conventional liquid sprays would not reach. Foam is created by adding a foaming agent to the finished dilution of insecticide, then mixing air with the liquid before it is sprayed. Combining air and liquid produces the foam.

- *Dry foam* consists of a large amount of air and a small amount of liquid.
- *Wet foam* has a large amount of liquid and a small amount of air.

SOLVENT-BASED FORMULATIONS

Solvent-based formulations consist of an active ingredient dissolved in light oil or organic solvent. Since the insecticide is soluble in these liquids, they are true solutions. Most solvent-based sprays are packaged as ready-to-use products. Oil-based formulations were commonly used to control indoor pests. However, their use has diminished with the development of more cost-effective, water-based formulations that do not have the disadvantage of being flammable, yet provide contact and residual control.

Fogging. The efficacy of indoor space sprays and outdoor fogging is influenced by droplet size, which is directly influenced by the oil or water base of the formulation. For droplets to be effective as a contact and/or surface treatment, they must retain a small size, of less than 50 microns, while airborne.

Shrinkage of droplets after application limits their utility for space sprays. Water-based droplets can remain airborne for more than 15 minutes, but they may become too small to impact a flying pest or produce a lethal residue on a surface.

Oils. Oil-solution concentrates have a technical-grade insecticide in solution in oil. Some oil-based formulations are ready to use, requiring no further dilution. Oil dilutions usually provide faster knockdown than other formulations. This may be due to the solvent having insecticidal qualities, and because the active ingredient can be carried through the insect cuticle in the solvent. However, oil-based formulations are absorbed by porous substrates, and this can limit the amount of insecticide available to pests walking on the surface.

Evaporation-Retarding. These formulations are characterized by the inclusion of long-chain alcohols. This decreases the amount of water evaporating from the small droplets in the spray. Evaporation occurs when generated droplets move through the air to the target pest or substrate. These formulations can provide good performance by maintaining a biologically effective droplet size.

Lacquer. Insecticide lacquers were developed for the control of cockroaches and other indoor crawling pests. They were formulated with high insecticide concentrations of 4% to 10%. The objective was to establish a residue that would remain active for an extended time. Lacquers were used to treat locations that sprays or dusts would not reach and to remain effective when exposed to household or commercial conditions. The incorporated insecticide is produced on the surface of the lacquer to provide a type of controlled-release delivery. Application is usually made with a brush to avoid contaminating non-target surfaces with a spray.

Insecticide concentration on the surface of lacquers gradually declines, eventually to a non-lethal level. Long-term exposure to sub-lethal concentrations can kill susceptible individuals in the population, leaving only the most resistant. The result may be a population that is difficult to control because of high insecticide resistance.

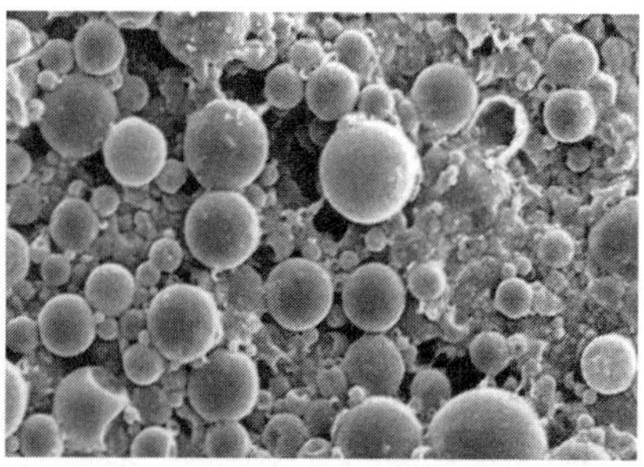

Microcapsules (through electron microscope)

CONTROLLED-RELEASE FORMULATIONS

The residual activity of an insecticide can be extended through its slow release from a treated surface or into a space, thus providing long-term control of crawling or flying pests.

Microcapsules. These formulations are also called microencapsulated (ME) and capsule suspensions (CS). They consist of liquid insecticide enclosed within organic spheres with diameters ranging from 2 to 50 microns. The objective is for these spheres to slowly release insecticide onto the treated surface. The speed of pest knockdown and the residual life are linked to the capsule wall; with diffusion, the thicker the wall, the longer the residual life of the application.

Insecticide is released from the capsules by either disruption (breaking) of or diffusion through the wall. Disruption releases the insecticide soon after application, but diffusion is controlled by the porosity and thickness of the capsule wall. Some formulations contain capsules with thick walls as well as capsules with thin walls. The thin-walled capsules provide knockdown of the target pest, while the thick-walled capsules diffuse the insecticide slowly to provide an extended residual. Capsules can also have dual-wall construction, which provides added strength and can ensure a release rate of up to 90 days.

Microencapsulated formulations offer the advantages of protection against breakdown in unfavorable conditions, long residual efficacy, and reduced volatility. The extended life of ME residue can result in immediate or delayed transfer of a toxic dose to the pest when capsules are contacted, ingested, or crushed by the tarsi of a walking insect. Fully intact capsules may be on the surface for weeks after application. The immediate and residual action of microencapsulated formulations is often controlled by the treated substrate; nonporous substrates are more favorable than porous ones.

When microcapsules are picked up by crawling insects, the liquid diffuses out and onto their cuticle to provide knockdown and control.

Microcapsule formulations are typically designed for liquid-spray delivery, but are also formulated as a dry foam (aerosol) delivery.

Small capsules in a formulation may settle within the surface matrix of common porous substrates and not be available to the pest, as the cracks and crevices in bare, unpainted wood may keep the capsules from contacting the tarsi of crawling insects. However, these substrates would increase the residual life of the application.

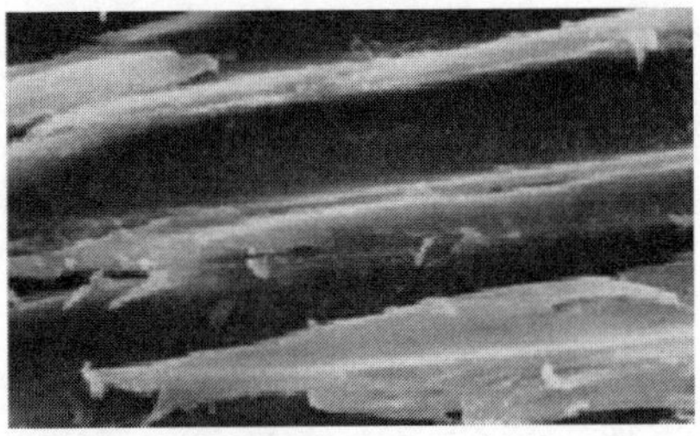

Bare wood (through electron microscope)

Plastic Panels and Strips. Controlled release of an insecticide can be achieved by incorporating a volatile compound in a plastic resin. Rates of insecticide loss at the surface are controlled by the plastic and the insecticide. In general, the rate of insecticide loss is nearly constant over a period of about four months. The effective concentration of insecticide in the treatment area may depend upon temperature and barometric pressure. It can take several hours to achieve a concentration lethal to common insect pests. Strips are designed to protect a specified volume of space for a specified time.

Dichlorvos (DDVP), a volatile organophosphate insecticide, is available as DDVP-impregnated resin plastic strips for household pest control. Small strips can be used continuously in residential closets and storage rooms. Large strips are limited for use in areas where human exposure is of less than 4 hours per 24 hours. Treatment sites may include attics, garages, crawlspaces, and outbuildings. The strips can be used continuously if the space is unoccupied for 4 months after installation.

Bed Nets. Modern bed nets are designed with controlled-release technology. The configuration of the fabric retains insecticide for release during the practical life of the net; originally, bed nets had to be periodically re-treated with insecticide. The insecticide acts when target mosquitoes fly near or land on the net. Plastic polymer nets incorporate permethrin into the fibers of the net. The combination of the chemical formulation and fabric structure provide exclusion and an effective residue for contact kill.

Treated Mattress Liners. The typical fabric encasements for a mattress or box spring act to prevent the movement of dust mite allergens, mites, and bed bug adults and nymphs. An insecticide-treated mattress liner can provide a degree of protection and control of bed bugs, primarily by limiting their movement and denying them egg-laying sites on the bed. Polyester liners are treated with permethrin, a pyrethroid insecticide, and the efficacy lasts about two years. Treated liners are installed on top of the mattress, and pulled down and under the mattress edge. The liner is usually covered with a mattress pad or bottom sheet of the bedding.

Mosquito Coils, Vaporizing Mats. Mosquito coils are a different form of controlled-release technology. The slow burning releases low concentrations of insecticide, usually a pyrethroid, into indoor and outdoor spaces. These products require a heat source to vaporize the active ingredient from a combustible substrate. The combustible base of the coil binds the insecticide until ignited and allows the coil to burn with a self-sustained ember. Mosquito coils produce a visible smoke when ignited; they may be used indoors and outdoors.

Vaporizing mats rely on an electric heater to stimulate release of insecticide from the impregnated substrate. Mats provide protection from mosquitoes without emitting visible smoke. Controlled-release systems for outdoor use include devices that rely on heat from a flame to release insecticide from an impregnated cellulose mat.

Ambient Vaporizers. Indoor and outdoor mosquito control can be provided by the controlled release of insecticide from vaporizers. Slow release is accomplished without heat by using specially designed dispensers or emanators that utilize room temperatures. The delivery mechanism is usually an impregnated paper substrate designed to expose multiple surfaces. Paper emanators rely on fans or natural air movement to extract and distribute the insecticide. The vapor-active pyrethroid metafluthrin is used in ambient vaporizer strips; this insecticide has a relatively high volatility.

Briquets. Briquets are large water-soluble granules that release effective levels of insecticide or insect growth regulator for about 30 days. This type of controlled- or sustained-release formulation is used primarily to control mosquito larvae in small bodies of standing water. In general, they can be applied prior to the formation of pools and standing water. They are unaffected by dry-down situations and are quickly re-activated when conditions become water soaked.

AEROSOL FORMULATIONS

Aerosol spray is an efficient way to deliver small amounts of liquid insecticide to a substrate or void space. Inside the typical aerosol can is the insecticide and a propellant. The combined ingredients, insecticide and propellant, are pressure-sealed then released by depression of a valve. In the past, chlorofluorocarbons (CFCs), including Freon, were used extensively as aerosol propellants. However, CFCs are linked to depletion of the earth's ozone layer and have been replaced. The most common replacements are mixtures of propane, n-butane and isobutene; dimethyl ether (DME) and methyl ethyl ether are also used.

Liquid Propellants. These are gases that exist as a liquid under pressure. Since the aerosol is under pressure, the propellant exists mainly as a liquid. As the product is sprayed from a can, some of the liquid propellant vaporizes to fill the head space. This maintains constant pressure in the can, enabling consistent spray delivery until empty.

Compressed Gas. These propellants occupy only the head space in the aerosol can. They are not mixed with the insecticide product like the liquid propellants. When the actuator is pressed, the gas pressure acts to push the liquid out of the can. The amount of gas in the can remains the same, but as the product is dispersed through use, the pressure in the can will drop. Since there is no solvent-type propellant that vaporizes when the spray emerges, only the product is sprayed. Small droplets are created through mechanical action; as the liquid passes through the actuator, it is broken up into small droplets.

SOLID FORMULATIONS

Gels, granules, and dusts contain both the active ingredient and an inert carrier. Granule formulations combined with a food lure may target foraging and feeding habits of pests. They combine a toxicant with a desirable food source and are a suitable size for transport by the target pest. Typical granular insecticides deliver the insecticide to the area immediately around the particle.

Dust. Dusts are ready-to-use formulations of a low, 1% to 5% concentration of insecticide mixed with a carrier, such as clay. Dust particles are easily picked up by crawling insects and may remain as an active residue for long periods. The transfer of the dust particles to the insect may be by contact of its legs on treated surfaces, or by grooming of legs, wings, and antennae that have become contaminated. Formulations of desiccant dusts may contain pyrethrins, which provide quick knockdown and act as a flushing agent. The increased toxicity of some dust formulations may be due to penetration of the toxicant after abrasion of the cuticle.

Dust particle size of about 5 microns is optimal for transfer to cockroach legs during its walking and grooming.

Chalk. Sticks of blackboard chalk infused with liquid insecticide have been used to control cockroaches and other crawling insect pests. This material is calcium sulfate (gypsum), and it absorbs water-based dilutions or solutions of insecticide. Common chalk readily crumbles when used, leaving particles that stick loosely to the substrate. Insects contacting the chalk line may pick up a lethal dose of insecticide on their legs or antennae. However, the safety of this formulation and application method is questionable.

Desiccants. Dust desiccants are natural or synthesized materials that have the ability to disrupt the wax coating in the cuticle of arthropods and bring about a lethal rate of water loss. Non-abrasive dusts can be effective if they are highly sorptive. Desiccants are toxic to nearly all insects. Dehydration caused by desiccant dusts does not provide pest knockdown, but movement declines gradually until death occurs. Because of their purity and chemical inertness, silica gels are not toxic to humans when inhaled. The silica gels with low bulk density and the greatest porosity are called aerogels.

Granules. The use of granular insecticides is primarily for outdoor pests, such as ants, field crickets, centipedes, and millipedes. Granules can be applied to bare soil, turfgrass, and mulch as a spot, limited-area treatment, or entire perimeter treatment. The active ingredient is attached to an inert clay or organic material that acts as a temporary carrier. The carrier eventually degrades or is incorporated into the soil. The delivery format may be released of the insecticide into the soil, and this release may be enhanced by moisture in the soil or in the thatch layer in turfgrass.

Granules may be designed to be transported to the nests of social insects, such as ants, and the insecticide distributed to immature stages and nestmates. The size of the granules targeted for ants may correspond to the size of the mouthparts. Some formulations are designed to be small to enhance penetration of mulch or turfgrass thatch. Light-colored granules can enable better monitoring during application and ensure uniform coverage. Some carriers act as soil conditioners to promote aeration of treated soil and can degrade to provide micronutrients to plants.

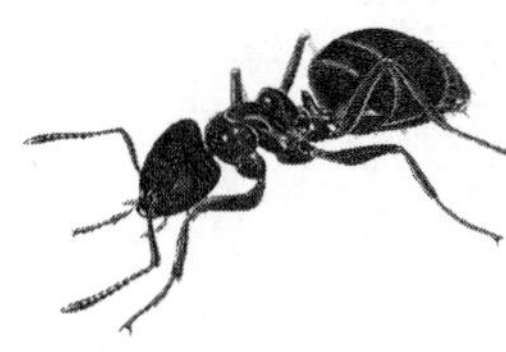

Ant

Granular size can be further enhanced to link the different size granules to specific combinations of food attractants for ant control. Some granules may be high in protein, some with a high percentage of simple sugars, fats, or complex carbohydrates. Ant species prefer different types of food, and preferences within a single species, and even within a single colony, can change slowly based on seasonal climate change or quickly based on competing food sources.

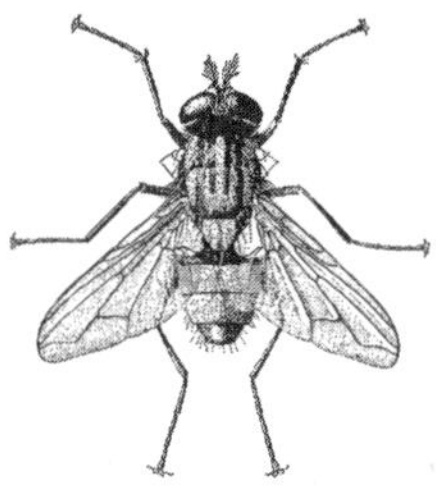

Granular bait has been also effective in controlling house flies in outdoor locations, especially in dumpster stalls where the bait can remain out of public contact.

Smoke. A smoke formulation consists of a mixture of an active ingredient with an organic fuel and an oxidant (typically sodium chlorate) that makes it burn. On ignition, the mixture generates a cloud of water vapor, carbon dioxide, carbon monoxide, and the vaporized insecticide. The insecticide condenses from the hot vapor to particles of 1 micron or less. There is some aggregation in the smoke cloud that results in slightly larger particles. The particles are carried by indoor air currents and dispersed over a wide area. Insects crawling over surfaces covered with the microscopic smoke particles pick up a lethal dose of insecticide.

A common misconception is that smoke acts as a fumigant, but solid particles of insecticide are produced. The particles are deposited on horizontal surfaces, with little penetrating into voids or sticking to vertical surfaces.

Baits. A variety of active ingredients, from borates to neonicotinoids, have been incorporated into insect-control baits. The consistency of baits includes liquid, solid (granular), semi-solid, paste, and gel. Moist baits containing 50% to 75% water are attractive to cockroaches and ants. These pests find their food by random searching, using their antennae and sensitive mouthparts to detect odors and moisture. They are naturally attracted to moist bait formulations.

Cockroach baits are often thought to contain an ingredient that attracts cockroaches. While there is usually a feeding stimulant in the matrix, this acts to increase feeding and not as an attractant. The high water content is what is attractive to the cockroaches.

Modern cockroach control baits usually contain a feeding stimulant, which ensures that, once the insect begins feeding on the bait matrix, it is likely to continue feeding and ingest a lethal dose. Glucose, fructose, sucrose, and maltose have a stimulant effect on the feeding of the German cockroach and probably other cockroach species.

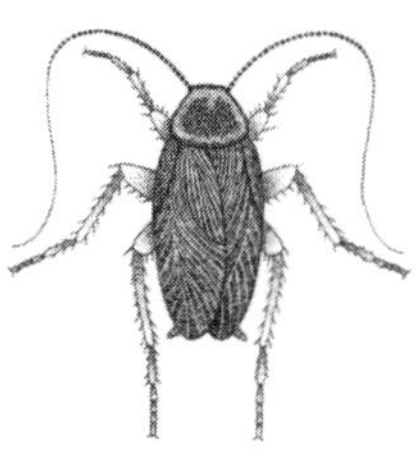

Gel bait formulations for cockroach control are influenced by environmental conditions, which usually results in a loss of moisture. Under standard household conditions of 65% relative humidity at 67°F, a gel bait placement can lose 43% of its moisture in the first 24 hours.

Bait formulations for ant control rely on the routine exchange of liquid food between nestmates. Concentrations are designed to be low and not lethal to the workers transferring bait. The insecticide in the bait accumulates in lethal quantities in larvae and male and female reproductives in the nest.

Edible dust formulations of baits are designed for termite control and are based on cellulose combined with an insecticide. The inert ingredient is food-grade cellulose, which is a natural food for termites. These dusts are capable of acting by direct contact utilizing the active ingredient, or by ingestion in the form of bait. The small particle size also permits the transfer of a lethal dose between individuals in the same colony; transfer ratios of 1:10 can be expected in some treatments.

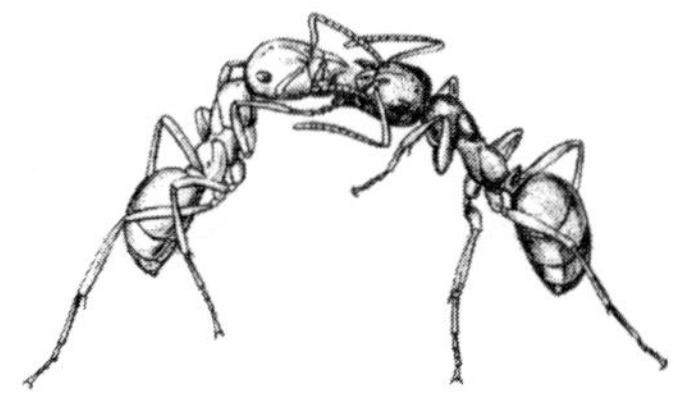

Ants sharing food

CHAPTER 6

APPLICATION TECHNOLOGY

The application of liquid, foam, dust, and granular insecticides requires precision equipment and a knowledgeable applicator. The technology behind the equipment is based on the mechanics of pressure, nozzle flow rate and spray pattern, and droplet size. These features are interrelated, and they influence the delivery, surface coverage, and performance of an insecticide.

LIQUID APPLICATION

The nozzle or spray tip on the valve extension of a sprayer may have an orifice that produces a coarse fan spray and one that produces a fine fan spray. Fan sprays differ by the angle of the fan, the size of the droplets in the spray, and the volume of liquid delivered per minute.

Nozzles. The nozzle controls the rate, pattern, and uniformity of a spray application. It is the most important part of application equipment, but typically the least understood and cared for. The term "nozzle" is commonly used for the spray tip at the end of the valve extension. It may spray from a single orifice opening or from several, such as a four-way nozzle, which has a fine and a coarse fan and a large and a small pin stream.

- *Flat fan spray.* These sprays are flattened and angled in the shape of a fan. The resulting spray pattern on the surface is a straight line, but slightly enlarged in the middle. The actual orifice has a flattened oval shape. The flat fan sprays used in household pest control have a 50° coarse fan or 80° fine fan spray angle. The optimum pressure for these nozzles is 20 to 40 pounds per square inch (psi). Above and below this pressure, there is a change in the liquid flow rate and droplet size. Low pressure reduces flow rate and increases droplet size, high pressure increases flow rate and decreases droplet size.

  The distribution of liquid in the fan spray is not uniform across the swath when it contacts the surface. The majority of the liquid is delivered at the middle of the swath, with less at the two ends. This spray pattern is maintained regardless of the pressure at the nozzle, but it is more pronounced at low pressures because the swath is reduced in length.

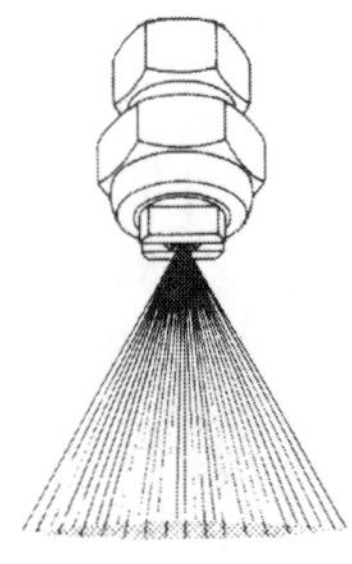

Fan spray

Fine fan sprays deliver 130 micron droplets, and coarse fan sprays deliver 240 micron droplets.

"Even" flat fan nozzles deliver uniform, even coverage of liquid across the spray pattern. The numbers for these nozzles usually include an E to indicate an even pattern.

- *Cone spray, pin-to-cone.* These nozzles produce a cone-shaped spray. The hollow cone-spray pattern is a circular ring of droplets; the full cone-spray pattern is circular and completely filled with spray droplets. The pin-to-cone nozzles are adjustable, and can deliver a pin-stream or hollow-cone spray.

  These nozzles are designed for agriculture and primarily used to deliver insecticide over the top of plants for complete coverage. A range of delivery rates, spray angles, and droplet sizes can be obtained from these nozzles. The typical spray angles are 65° and 85°.

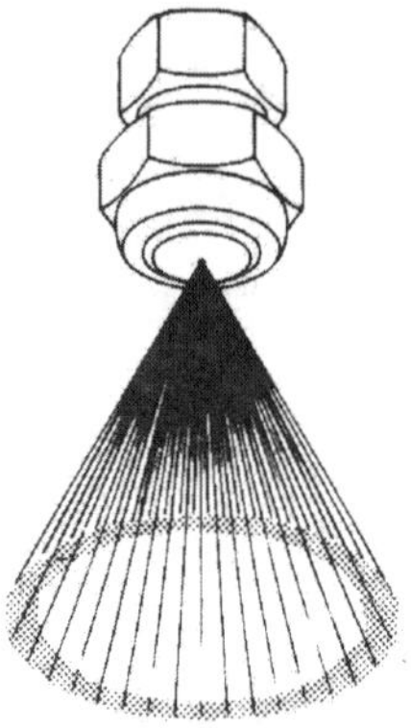

Cone spray

Cone and pin-to-cone nozzles are not suitable for perimeter applications to building foundations. These nozzles are not designed to treat flat surfaces; the flow rate is higher than a coarse fan spray.

- *Pin stream.* This is a narrow stream of liquid. There is no spray angle or pattern to the liquid delivered, and no distinct droplets. The actual orifice is round and the diameter determines the delivery rate. There is limited use for a pin-stream spray in indoor applications, as it floods the treated surface with liquid, which usually results in run-off. The pin-stream spray previously used for crack-and-crevice application has been replaced with metal or plastic straws. These tips can be inserted into the crevice to increase efficacy and decrease exposure to insecticide.

Spray pattern

Crack-and-crevice straw

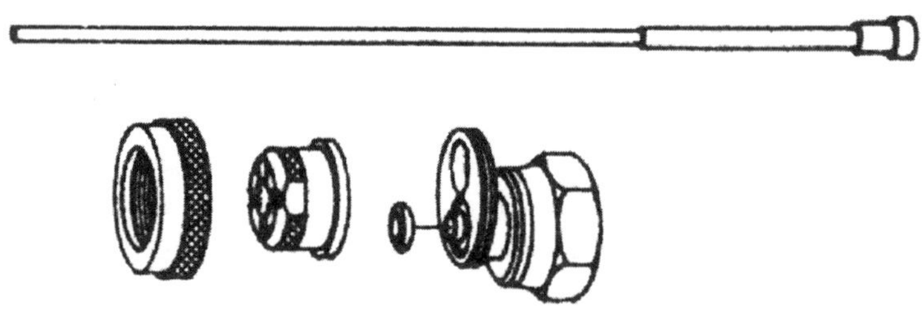

- *Crack-and-crevice straws.* These are extensions of a pin-stream orifice; they deliver a straight stream of liquid. The actual "straw" can be plastic or stainless steel. The use of a crack-and-crevice straw enables application of the liquid insecticide inside a harborage with little or no contamination of non-target surfaces.

- *Foam nozzles.* These nozzles draw air into the spray liquid to create foam. Introduction of air and the addition of a surfactant, a foaming agent, into the liquid insecticide results in production of foam bubbles. Foam delivery systems produce foam that quickly breaks down to deposit liquid insecticide on the substrate. Dry foam delivers only a small amount of liquid and persists for a long time. The advantages of foam application are its spread over uneven surfaces and fill of voids. Foam application is commonly used in treating the void beneath concrete slabs and infested wall voids.

50 micron droplets

- *Cold fogger nozzles.* Droplets can be created by impacting a fluid containing an insecticide with a rapidly moving air stream. The flow rate of the air and liquid can be adjusted to influence the size and delivery of droplets. Large droplets are produced with increased liquid flow and a reduction in air velocity; small droplets are produced by increasing or maintaining air velocity and decreasing liquid flow.

- *Thermal fogger nozzles.* These nozzles are used to apply insecticides in small droplets or fog. Insecticide is dissolved in diesel oil, kerosene, or other oil that has a high flash point. The mixture of insecticide and oil is injected into a stream of hot gas, which breaks the liquid into droplets, which then vaporize. The vapor is carried into cool outside air and condenses into a dense cloud. The cloud or fog contains so many droplets below 15 microns that visibility is reduced.

Nozzle Identification Number. Nozzles are assigned numbers that indicate flow rate and spray angle. The first two numbers in the sequence indicate the angle of the spray when it is discharged from the orifice. The remaining numbers indicate the gallons per minute (gpm) delivered at a specific pressure, which is typically 40 psi.

- *800067* indicates that the nozzle has an 80° fan angle and a flow rate of 0.067 gpm (8.5 ounces/minute).
- *50015* indicates that the nozzle has an 50° fan angle and a flow rate of 0.15 gpm (17 ounces/minute).
- *8002* indicates that the nozzle has an 80° fan angle and a flow rate of 0.2 gpm (2.6 ounces/minute).

8002 nozzle

Nozzle Flow Rate. The amount of liquid that passes through the nozzle is the flow rate; it varies with the pressure in the delivery system. To double the flow rate through a nozzle, the pressure in the system must be increased four times. Higher pressure not only increases the flow rate, it also influences the rate of orifice wear—as pressure is increased, the rate of orifice wear increases.

Spray Angle and Swath Coverage. The coverage of liquid on a treated surface is the spray swath, or band. For a flat fan spray the swath is lens-shaped; it is about 7 inches wide and 20 inches long. The enlarged center of the swath is due to the concentration of liquid in the middle of the pattern. There is less liquid at the two edges of the swath than there is in the center. The length of the swath depends on the spray angle and the distance the nozzle is held from the substrate.

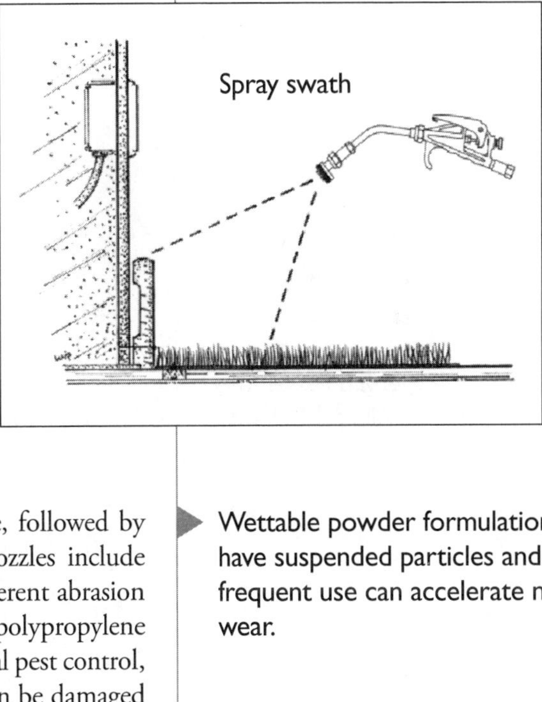

- *An 80° nozzle* held 18 inches from a surface has a swath of 30 inches long; held 24 inches from the surface, the swath is 40 inches long.
- *A 50° nozzle* held 18 inches from a surface has a swath of 17 inches long; held 24 inches from the surface, the swath is 22 inches long.

Nozzle Wear. Nozzle wear may be evident as an increase in flow rate, followed by a deterioration of the spray pattern. The composition materials for nozzles include ceramic, stainless steel, polypropylene, and brass. Each of these has a different abrasion resistance number. For example, the resistance number for brass is 1, for polypropylene 1 to 2, and for stainless steel 4 to 6. Brass nozzles are economical for general pest control, but they have limited service life. They are susceptible to abrasion and can be damaged when improperly cleaned.

- Wettable powder formulations have suspended particles and their frequent use can accelerate nozzle wear.
- After 300 hours of normal use, the flow rate for brass nozzles can increase by as much as 27%.

Droplet Characteristics. Liquid sprays are classified by size of droplets in the spray pattern.

Aerosol sprays are composed of droplets larger than 50 microns. Sprays of smaller droplets are not visible to the human eye. The cloud of small droplets, of less than 50 microns, from a thermal fogger is visible because of the large number or density of the droplets.

- The term *"fog"* is usually applied to an aerosol spray with droplets in the range of 5 to 15 microns.
- *Mists* are composed of droplets that are 51 to 100 microns.
- *Fan sprays* can be composed of fine (101 to 200 microns), medium (201 to 300 microns), and coarse (less than 300 microns) droplets, depending on the nozzle.

Droplet Size. The size of droplets produced by a fogger is often used as an indication of performance. The use of a single number in product specifications creates the misconception that all the droplets in the spray are this size. However, spray nozzles always produce a range of droplet sizes. Sprayer performance can be misrepresented by stating the smallest droplet in the range. These droplets are usually few in number and do not carry an amount of insecticide that is actually lethal to the target pest.

Volume median diameter

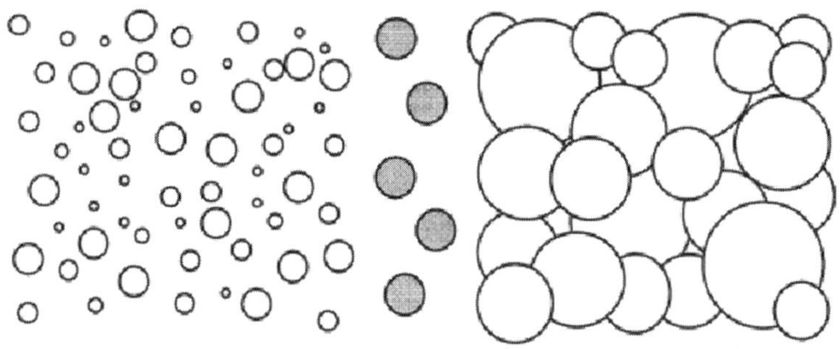

It is a common misconception that the VMD droplet size produced by a cold or thermal fogger is the average size of the droplets produced. It is not.

- *Volume Median Diameter (VMD).* The most commonly used measurement for droplets is the volume median diameter (VMD); it is also expressed as $D_{v0.5}$ and as mass median diameter (MMD). The VMD number is the median size of the droplets, those in the middle of the range of all the droplet sizes produced. The range can extend from very small (less than 10 microns) to very large (more than 300 microns), depending on the nozzle. Regardless of the range in size, the VMD is used to characterize the spray.

  The VMD divides the spray into two equal parts by *volume*. The two halves of the spray have the same total volume of liquid, but the size of the droplets in the two halves differ. One half contains droplets that are smaller than the VMD; the other half contains droplets larger than the VMD.

  The VMD represents only about 10% of the total number of droplets in the spray; 45% of the droplets are smaller than the VMD, and 45% are larger than the VMD. The VMD droplet size is not the smallest and not the average droplet size produced by a fogger.

- *Number Median Diameter (NMD).* The NMD value is commonly used in the marketing of foggers because this value is usually smaller than the VMD. But the NMD is not an accurate indication of the size range of droplets delivered by a fogger.

  The NMD divides spray droplets into two equal parts based only on the total *number* of droplets. Typically there are more small droplets than large droplets in a spray, and the NMD places emphasis on the small droplets. It is not more accurate than the VMD in characterizing the droplets produced by a fogger.

  If all the droplets in a spray are arranged in a series from smallest to largest by

Number median diameter

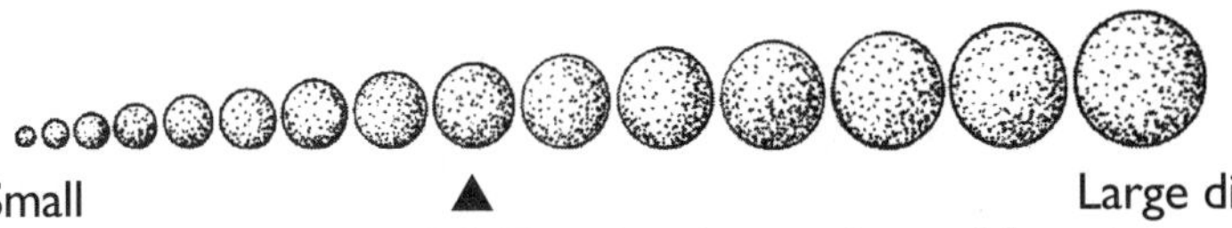

diameter, the diameter of the droplet at the mid-point of the series would be the NMD. For example, if there were 15 droplets in a spray, 7 would be larger than and 7 would be smaller than the NMD.

Droplet Movement. The droplets close to the nozzle move in the air-stream produced by the spray and can be directed to a surface. Further away from the nozzle the air stream is weaker and air currents can carry small droplets away from the surface. This movement is called drift and can result in non-target exposure to the insecticide.

Usually large spray droplets of more than 50 microns are not moved by air currents, but fall quickly to the surface. Droplets of less than 30 microns may take several minutes to fall, and during this time they are subject to movement by air currents. The distance a droplet travels depends on the strength of the surrounding air currents, whether it remains the same size, and the obstacles in its path.

- *Evaporation.* Small droplets have a large surface area but small volume. Through evaporation, a small droplet will lose volume over its surface and become smaller. As they fall through the air, small droplets in a water-based spray decrease rapidly in size.

- *Electrostatics.* Droplets can be made to move toward a surface when given a small electric charge. The maximum electric charge a droplet can carry depends on the size of the droplet, and, to some extent, on surrounding air. Electrostatic spraying is most effective when droplets are small (30 microns) and the electric charge is relatively high. High humidity in the air can reduce the droplet charge and decrease the electrostatic attraction to the substrate. A strong air-stream is usually necessary to place the droplets close to the target substrate. Electrostatic droplets must be close to the oppositely charged substrate to be attracted.

- *Ultra-low volume (ULV).* The terms used to describe the amount of liquid in a spray are: high volume (HV), very low volume (VLV), and ultra-low volume (ULV). Ultra-low volume is defined as the minimum volume per unit area of insecticide necessary to achieve control. It is an application method and not a specific droplet size.

  ULV is commonly used to identify a small droplet size, but there is no specific droplet size for a ULV spray. The actual size of the droplets in the ULV spray depends on the size of the target pest. For example, the ULV droplet size for mosquitoes is typically 17 microns, but this would not be the same for the much larger meal moths, for which droplet size is typically 30 microns.

Surface Contact and Coverage. Target sites for indoor and outdoor spraying vary from smooth artificial surfaces to rough natural surfaces. Surface roughness is important to the retention of spray liquid after application. Roughness can be responsible for reducing the potential of droplet bounce and increasing droplet capture. Since the efficacy of spray treatments is usually linked to the amount of liquid applied, it is useful to characterize spray coverage as high, medium, and low volume.

- *High volume.* The substrate is treated until surface saturation occurs and excess spray liquid runs off. Insecticide labels that permit high-volume coverage usually prescribe spraying only to the point of run-off.

- *Medium volume.* The treated surface has distinct drops. A dense pattern of small drops leaves little space between droplet residue for crawling insects to miss or avoid while walking on the treated surface.

- *Low volume.* The spray is deposited as distinct droplets, and their size and spacing is variable. The result is a random pattern of dried droplet residue on the treated surface.

Run-Off. During spraying, drops on the surface begin to overlap and join at their edges. Depending on the surface and the amount of insecticide applied, large groups of joined droplets can collect on the surface. Eventually, one of these enlarged drops becomes unstable and rolls down the surface. This may trigger other large droplets to move as it travels past them or into them. Further spraying causes additional droplets to coalesce and reach the size needed to roll down the surface. This movement is run-off, and it can result in reduced surface coverage, regardless of how much was applied.

- *Droplet size and density.* A reduction of the droplet size in the spray generally increases retention of droplets on the surface and reduces run-off. This is expected, since small droplets generally produce an even pattern on the surface. Fine sprays provide a significant increase in retention over coarse sprays.

- *Wood surface characteristics.* Liquid application to untreated wood is complicated by rough and smooth areas, knots and checks, and vertical and horizontal edges. Structural timber may have a raised grain and a rough surface, or a smooth surface. The rougher the surface the less likely run-off will occur. Vertical edges may not retain droplets or they may be shielded from the spray pattern. Horizontal edges may receive increased amounts of liquid due to gravity flow or run-off.

AEROSOL APPLICATION

Aerosol Sprays are composed of droplets that are less than 50 microns (VMD), the most effective size range is 20 to 50 microns diameter. Droplets larger than 50 microns fall out of the air quickly and are not effective. Droplets of less than 20 microns may not carry a lethal dose of insecticide for most household pests.

Voids and small spaces are more successfully treated with a direct power mist, which has droplets larger than 50 microns; these droplets carry a lethal dose of insecticide and result in an effective residue.

Aerosol space sprays can be used to treat confined spaces with droplets that remain in the air for long periods: 20 micron droplets will stay airborne for 4.2 minutes; 50 micron droplets, for 40.5 seconds.

The objective may be to control flying insects, to flush insects from harborages, or to create a droplet residue. However, the ability of small droplets to penetrate dead-end cracks and voids in which there is no air circulation is limited. In aerosol sprays, 95% of the droplets settle on horizontal surfaces, 5% on walls and ceilings.

Aerosol Can Systems. This delivery system provides maximum efficacy in the application of a small amount of insecticide to a specific location, such as a pest harborage. Insecticide residues can remain active and available to the target pest for long periods, and there is little or no exposure to people or pets following the application. Aerosol delivery now includes dust formulations for crack-and-crevice (C+C) or surface application.

In the standard aerosol-can delivery system, the container and the nozzle are combined into one unit. The shape and size of the orifice in the actuator determine the spray pattern, droplet size, and delivery rate. It may be designed to hold a C+C straw or other type of void tip, or to produce a unique spray, such as a power pin-stream or foam.

Modern aerosol application extends the spray tip from the top of the can through a coiled hose and a hand-held valve. There is no loss in pressure from the can to the valve, and delivery is improved with standard spray and C+C application.

FOAM APPLICATION

Foam is typically used to apply insecticide to uneven or difficult-to-reach surfaces, including vertical surfaces. Foam can flow like a liquid, remain in place like a solid, and disappear in a short period of time. It is a material in which a gas, usually air, is dispersed in a matrix of bubbles that are separated from each other by a thin film of liquid. The liquid film drains to the surface, and the bubbles becomes thinner and finally collapse.

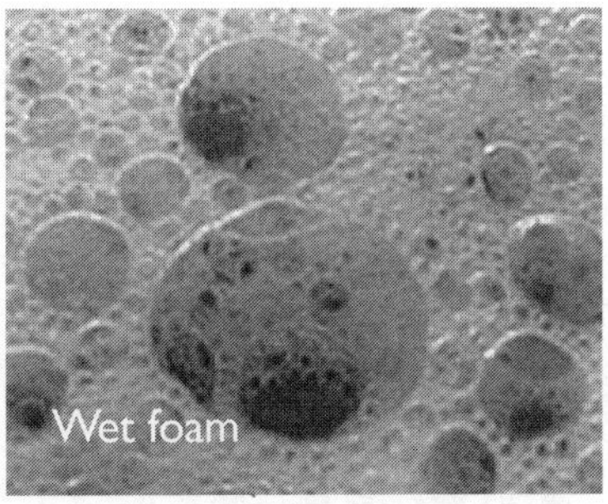
Wet foam

Composition. Foams are typically 95% air and 5% liquid. The liquid content consists of soap surfactants and insecticide active ingredient. The amount of surfactant controls the amount of foaming that takes place in a liquid. The soap surfactants reduce the surface tension, which permits the liquid to accept the air that will produce the foam.

The addition of a foaming agent to an insecticide should begin with the lowest recommended amount, which is typically 1 to 2 ounces of foaming agent per finished gallon of insecticide. Increase the amount of foaming agent only as needed.

Note: There is a limit to the solubility of surfactant molecules in a liquid. Once the limit is reached, the surface tension can no longer be decreased. The continued addition of surfactant to the original liquid will not increase the amount of foam. Adding more foaming agent does not continue to increase the amount of foam or the expansion ratio of liquid to foam.

Expansion Ratio. The expansion ratio is the amount of foam produced by a specific volume of liquid; ratios can range from 5:1 to 35:1. The liquid-to-foam ratio is determined by the equipment, the insecticide formulation, and the foaming agent. Foam application equipment is specifically designed to introduce a certain amount of air into the liquid stream, and to move the foam through the delivery hose to the nozzle. Changing any part, such as the length or diameter of hose, valve, or nozzle will likely change the foam.

Because of the active ingredient and solvent system of modern insecticides, some will produce foam more readily than others. This can not be predicted, but must be determined on a product-by-product basis. Similarly, commercial foaming agents have a specific combination of ingredients, each product is distinct, and each may produce more or less foam with a particular insecticide.

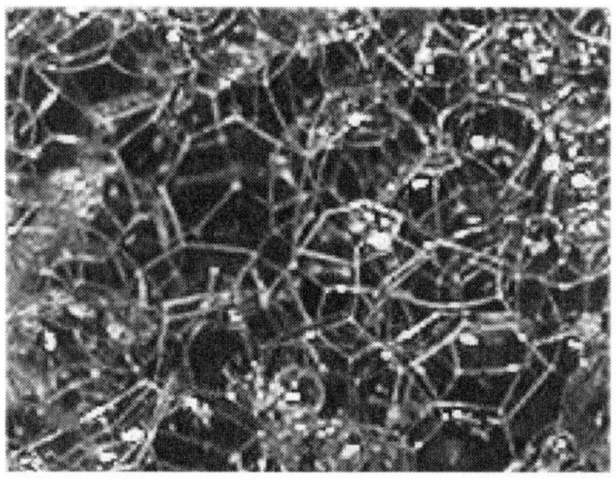
Dry foam

Wet Foam and Dry Foam. The combination and quantity of air and surfactant in a liquid produce two types: "wet" foams and "dry" foams. In wet foam the liquid content is high, and the walls of the individual bubbles are thick and have a spherical shape. In a dry foam the liquid content is low and the walls of the bubbles are thin and have an angular shape. As gravitational forces drain the liquid from the walls of the bubbles, wet foams become dry foams and eventually collapse.

Dry foam is generally considered a foam-to-liquid ratio of 25:1 or more.

DUST APPLICATION

Insecticide dusts are effective against crawling pests when applied as a light or thin covering. Proper application results in no visible residue. The objective of applying dust is to create a residue of insecticide particles on the target surface. The target pest will pick up a lethal quantity of particles by walking on the surface. Dust particles on legs and antennae are transferred to other parts of the body and to the mouthparts during grooming.

Excess amounts of dust applied to a surface may be repellent to some crawling insects, such as cockroaches, or may dislodge and contaminate other surfaces.

Insects generally avoid surfaces covered with loose material, such as dust or powder. The objective of applying dust is to create a thin surface layer that is too thin to be detected, yet thick enough to deliver a lethal quantity to the insects' legs.

Nozzles. Delivering a light coating of dust to narrow harborages and voids, without leaving a visible residue on the outside, requires extension tubes and directional tip nozzles. For hand- and power-operated dusters, these tips can help to meter the flow of dust. The C+C tip for hand dusters is a plastic or metal, tapered tube with a narrow opening at the end. This tip is usually fitted to the end of one or more extension tubes. The curved or 90° directional tips of hand dusters provide for a light covering of dust because the reservoir can remain horizontal.

Plastic C+C tubes are required for treating electrical outlets and other locations where the tip may contact active electric wires.

Electrostatics. In electric dusters the particles move rapidly through a plastic extension tube before exiting at the tip and may become electrically charged. The phenomenon of air-friction charging can give a negative (-) charge to the dust particles. The charge does not last long, but it can improve the spread of exiting particles over treated surfaces. The charge on the dust particle quickly dissipates once the particle contacts the surface.

For electrostatic "attraction" to occur, the target surface must be grounded and have a charge. However, many indoor surfaces do not have or do not maintain a charge, primarily because of the lack of grounding.

GRANULE APPLICATION

Granular formulations are available for baits, but most function as contact insecticides. The latter are applied as a broadcast treatment to soil, in crawlspaces, and as a band around building perimeters to control crawling insects, such as ants. The objective of contact-control formulations is to ensure an encounter between the granule and the crawling target pest within the residual life of the insecticide. Water-dispersible and other granular formulations bind to the soil after exposure to moisture, and function as a limited soil treatment.

Some granular bait products can be applied indoors. In food areas of restaurants and food plants, this may be limited to use as crack-and-crevice applications or in bait stations.

CHAPTER 7

TRANSFER OF A TOXIC DOSE

For insecticide residue to be effective, a small amount of the active ingredient must be transferred to the target pest. The lethal amount required depends on the insecticide and the target pest. Transfer of the toxicant to the pest may involve its walking on a treated surface; direct contact with droplets; or indirect contact through feeding, grooming of body parts, or exchange among individuals.

Transfer of liquid insecticides begins with the delivery of spray droplets and the formation of a residue. A microcapsule deposit consists of an active ingredient enclosed in a porous capsule. Fogging produces a curtain of droplets with sizes and fall rates to maximize contact with flying pests. Dusts are applied as a thin film on surfaces. Toxic baits depend on foraging/feeding behavior and food sharing between nestmates.

INSECTICIDE AND PEST INTERACTION

The three components of the insecticide and pest interaction are the pest and its behavior, the insecticide and formulation, and the substrate. The movement habits and foraging behavior of pests determine their encounter with spray droplets or a dry residue. The insecticide formulation (WP, EC, ME) determines the residue, the means of transfer, and residual activity. The substrate is important, since it is the intermediate between the delivery system and the target pest.

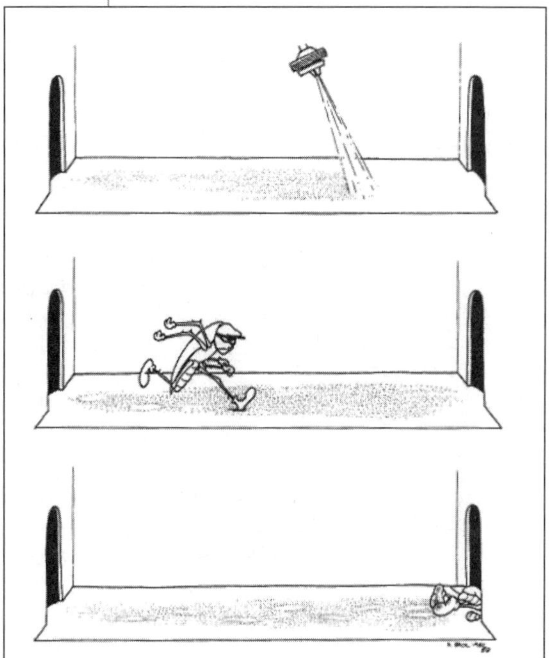

Pest. The amount of pest exposure to insecticides depends on the body part making contact with the residue. Tarsal pads on the feet are the primary contact point for crawling insects. Grooming moves particles and microcapsules from legs and antennae to the mouthparts. Consumption of toxic baits depends on the palatability of the bait and the potential for aversion to one of the ingredients.

Pests may avoid treated surfaces because their antennae can detect a volatile insecticide or an irritating component of the formulation. As a result, there may be increased activity near insecticide residues, with individuals moving away from a treated surface into untreated areas. Avoidance may be overcome by the attraction to food or a favorable harborage.

Formulation. The surface residue of powder, crystals, or microcapsules is determined by the formulation, the application nozzle, and the droplet size and spray distribution. Formulations can be designed to maximize residual activity and the transfer of a toxic dose to the target pest. Dusts may be transferred easily to crawling pests, but often need protection from water and humidity. High moisture content of gel baits increases initial attraction to the target pest. Wettable powder and suspension concentrates may leave behind solid residues, which are easily dislodged and transferred to a crawling pest.

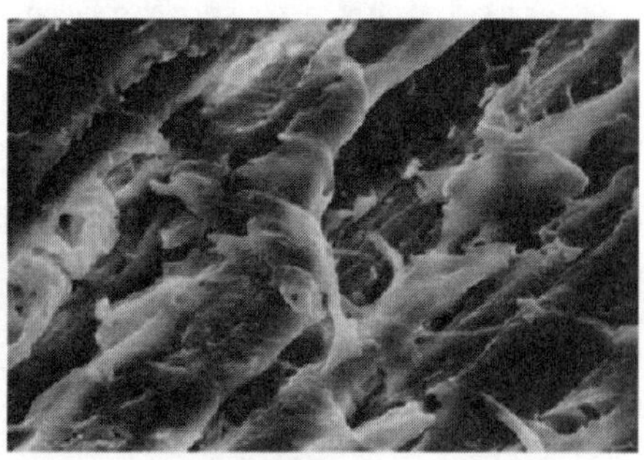

Latex painted surface

Substrate. The affinity of a droplet or dust particle for a treated surface influences adhesion after impact and transfer to the pest. Surface tension of the insecticide liquid is the main barrier to droplet spread on substrates. Increased surface tension prevents droplets from spreading and sticking on surfaces.

The capacity of a droplet to stick to a substrate is influenced by its size and the position of the surface. Horizontal surfaces are best treated with a medium or coarse spray. Droplets in these sprays are 200 to 300 microns in diameter and have the greatest spread and retention. Large droplets may bounce off the surface and fragment into smaller droplets. This is more likely to occur on a smooth surface than a rough surface.

RESIDUE TRANSFER: CRAWLING INSECTS

The toxic effect of surface residue is the result of the pick up of insecticide from the substrate. This occurs when insects walk across a treated surface and dislodge insecticide residue with the spines and scales on the tarsi and tarsal pads of their legs, or collect insecticide dust that adheres to their legs. Insecticide powder, crystals, or microcapsules readily adhere to the legs, antennae, and body.

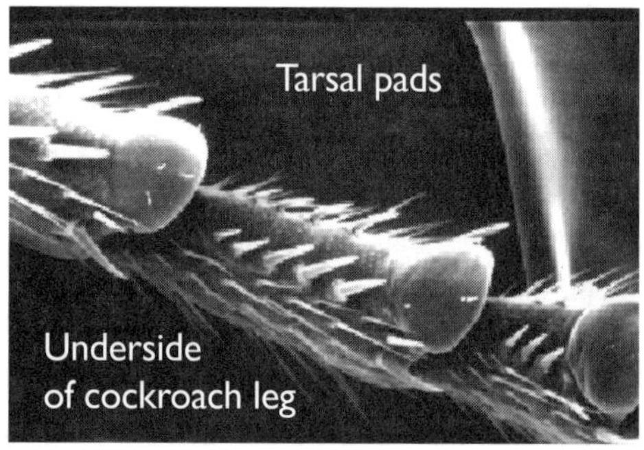

Insecticide dust and microcapsules smaller than 20 microns are picked up more readily and are less likely to be detached than are larger particles.

Weight and Microcapsules. The weight of the insect and the size of the tarsal pads can influence the transfer of a toxic dose; large insects usually have large tarsal pads and can easily acquire a toxic dose.

The capsules in ME formulations can be crushed by insects crawling on the surface. Capsules of 50-micron diameter and 0.14-micron-thick walls are crushed by the weight of adult German cockroaches. The average body weight of male German cockroaches is 51 mg, females average 53 mg, and females with egg cases average 111 mg. This weight is sufficient to crush capsules of most ME formulations, especially when contacted by the front legs. The average body weight of large nymphs is 44 mg; this may be enough to crush small capsules, but the weight of small nymphs may not be sufficient.

The weight of German cockroaches is centered on the front legs; it is the front legs that will first contact and crush microcapsules.

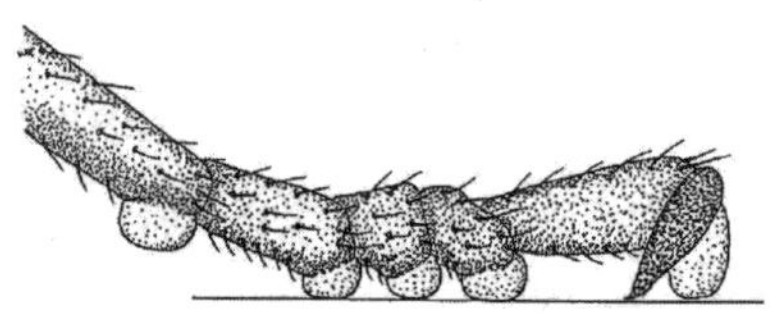

Tarsal pad contact surface

Adults and nymphs of the American and oriental cockroach easily crush microcapsules, but a large dose from numerous capsules is needed to be lethal.

When walking on a smooth surface, at least three of the five tarsal pads on the legs of cockroaches are used; the large hooked claws at the end of the leg are not used. When running on a rough surface, the last tarsal segment is bent down to engage the claw.

Walking and Running. The adult and immature stages of crawling insects have only their tarsal pads in contact with the surface when they are walking and running. Most flying insects, such as house flies and mosquitoes, divide their time between flying and resting on surfaces, and may not walk enough on a treated surface to acquire a lethal dose. Crawling insects, such as cockroaches, beetles, and ants, remain in continuous contact with the surface, and their tarsal pads provide large surface areas of contact.

- *Legs.* Cockroaches walk and run by coordinating their three sets of legs. The role played by each leg is different. Hind legs are used primarily for pushing forward, and they have large tarsal pads. This set of legs has a limited range of movement and carries the least amount of body weight. The middle legs are the fulcrum of the body. They move only slightly back and forth and have moderate-sized tarsal pads. The front legs are the most important when it comes to movement. They have a different pattern of nerves, move the greatest distance, and carry the most weight.

Underside of cockroach leg

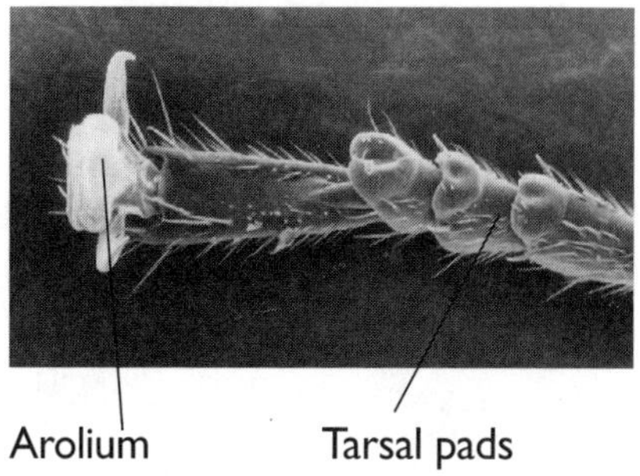

- *Tarsal pads.* The pads at the ends of the legs are involved in the pick up and transfer of insecticide residue from a treated surface. The tarsal pads are needed when walking on flat or inclined surfaces. The terminal leg segment, which bears the claws and the large pad (arolium) between them, is used when climbing vertical surfaces and walking upside down.

  In the oriental cockroach, the arolium is poorly developed; females and males have difficulty climbing smooth surfaces, and individuals are often trapped in sinks and tubs.

In the American and German cockroaches, the tarsal pads are present on all leg segments; they are large and rimmed with spines and setae. In the brown-banded cockroach only the last tarsal segment has a large tarsal pad. There is only limited transfer of insecticide residue to the legs of this species, making it more difficult to control with liquid insecticides and dusts.

The small size of the tarsal pads on the brown-banded cockroach may contribute to the low number of adults and nymphs captured on sticky traps. Individuals may be able to pull away stuck legs and escape.

- *Speed.* During standing and walking at low speeds, most of the legs of a crawling insect are contacting the surface. During fast walking and running, only the tarsi of three legs are in contact at one time. During walking, a cockroach creates both vertical and horizontal forces on the legs. Accelerating generates lateral forces that push the cockroach to the side. This lateral movement, along with the spines on the legs and tarsal pads, dislodges residue from the substrate.

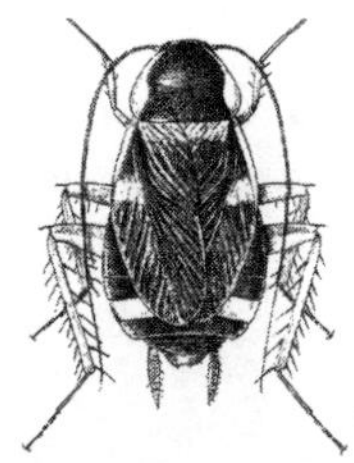

Brown-banded cockroach

Self Grooming. Grooming is an instinctive behavior, and insects regularly groom themselves. The primary function is cleaning of the sensory organs that detect food, pheromones, and volatile insecticides. This behavior redistributes particles of insecticides picked up from the substrate by the antennae and the tarsal pads of the legs.

Oral cleaning of the legs and antennae is common in cockroaches. It requires coordinated movements of the head and legs to bend the antenna to the mouthparts, and to extend the fore legs to the mouthparts. The front legs are groomed through the mouthparts, followed by the mid-legs. Particles of insecticides and microcapsules are removed from the antenna and the tarsal segments after one or more passes through the mouthparts.

RESIDUE TRANSFER: WOOD-BORING INSECTS

Wood-infesting beetle larvae contact treated surfaces when they bore below the surface of wood sprayed with liquid insecticide. The legs, mouthparts, and body segments of powderpost beetles and longhorned beetles are in direct and nearly constant contact with the surface of the gallery. The thorax and abdominal segments are pressed against the walls of the gallery. This gives the larva the firm grasp that allows their mandibles to gnaw the surrounding wood.

Larval Feeding. The large surface area of the larvae and the contact with the walls of the gallery increases the potential for insecticide penetration of the cuticle. Surface application of liquid insecticides to structural wood results in low concentrations immediately below the surface, but this may be sufficient for control. Old house borer larvae in contact with a surface treated with permethrin insecticide are killed by concentrations as low as 1 ppm (33% mortality) to 10 ppm (100% mortality).

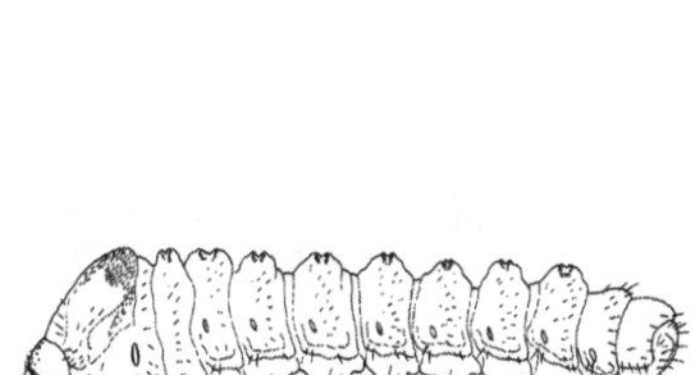

Wood-infesting beetle larva

Larvae usually rotate their bodies in the gallery to alternate the mandible gnawing the surface of the wood. This ensures exposure to uneven residues in the wood due to differences in penetration of liquids.

Tunnels. Larval feeding tunnels are usually several inches below the wood surface. They may be unaffected by surface application of liquid insecticide, which penetrates only 2 to 3 mm in most softwood timber. When larvae complete development they bore close to the surface to construct the pupal chamber. Pupal chambers are formed slightly below the surface; there is lethal contact with the insecticide residue at this outer layer of the treated wood.

During movement to the surface, larvae may simply bore and not feed on the wood. This may limit their exposure to some insecticides.

RESIDUE TRANSFER: DUSTS AND DESICCANTS

The transfer of a toxic dose of dust depends on the amount of insecticide on the treated surface, and the evenness of the layer on that surface. A uniform and small amount is optimal; a heavily coated surface is often repellent. Insecticide transfer and efficacy also depends on the formulation. Dust particles that are less than 20 microns readily cling to the legs, antennae, and body of insects and other arthropods.

Rough or porous surfaces reduce the amount of dust picked up by insects walking on these surfaces.

Wax Layer. Desiccants, such as silica, are materials that have the ability to abrade the surface or remove part of the wax layer from the cuticle. The wax can be removed by abrasion if the arthropod walks or crawls over abrasive particles. To be effective, abrasive particles must have a hardness number above 6.0 in the Mohs scale (a measure of mineral hardness). The result of exposure to these fine dusts is the loss of body water from the insect. Dehydration results in a decline in activity until death occurs.

German cockroach females carrying egg cases pick up greater amounts of dust than large nymphs, which, in turn, pick up more dust than males. Adults and nymphs of this cockroach demonstrate an increase in grooming when exposed to silica dust.

AIRBORNE TRANSFER: FOGGING AND MISTING

The intent of most fogging applications is to control flying insects, such as house flies and mosquitoes. Flying insects spend little time resting on surfaces, so knockdown and kill is achieved by treating the airspace in which they are moving. The substrate (air) can be treated with an aerosol or fog to create a temporary residue consisting of insecticide droplets.

Droplet Size. The basic objective of transferring a lethal dose to the target pest is often overlooked or misunderstood in fogging. Selecting equipment or delivery settings that produce small droplets ignores the fact that these are most likely non-lethal. Small droplets are easily carried away from the target pest or substrate by indoor air currents that move along the surface of floors, walls, and ceilings.

Space Sprays. The toxic effects of airborne sprays on a flying insect result from contact with suspended or falling droplets. The number and size of the drops in the spray depend on the delivery system. Insects flying in a space spray are moving through a treated substrate; in this case air is the substrate with suspended or falling droplets of insecticide. The number of drops contacted depends on the distance traveled in the treated air.

Harborages. Space sprays and room fogging are often used to treat cracks, crevices, and narrow harborages. The intent may be to flush insects, such as cockroaches, from harborages or provide contact kill with the spray droplet. This is an inappropriate and most likely ineffective application of insecticide. Large droplets in the spray fall quickly and are not carried into harborages. Small droplets can remain in the air for an extended time, but are not likely to move into harborages.

Fogging. Aerosol sprays are composed primarily of droplets of less than 50 microns and have a suspension time of about 40 seconds. Fog droplets have a diameter of 15 to 20 microns. These droplets remain in the air for an extended time, but when too large a volume is applied to a space, the small droplets combine to form large droplets, and these fall quickly.

Flying insects encounter drops because their wings sweep through large volumes of air, and small droplets impact their body. Large drops fall to the substrate. Small drops do not carry a lethal dose of insecticide, and can be carried around the insect in the air stream "boundary layer" generated by flight.

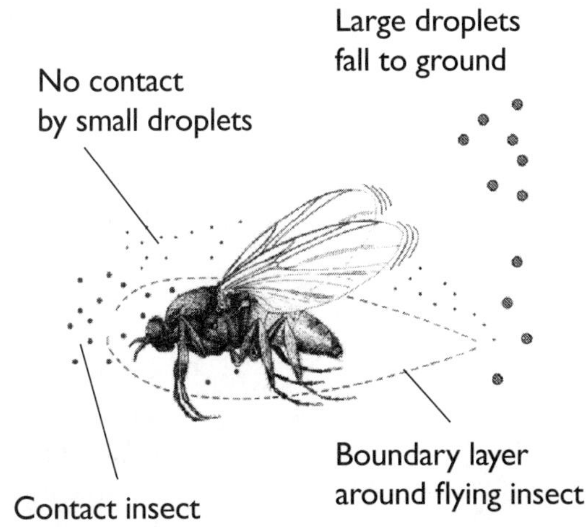

Power Misting. Mist blowers produce droplets in the range of 51 to 100 microns, and the droplets are delivered in a strong air stream. The optimum droplet size range for direct impaction on insects in flight or at rest is 15 to 30 microns. Drops larger than 50 microns are required for deposit on vegetation or other substrates. Power mist application for outdoor mosquito control directs the spray to the underside of the leaves on trees and shrubs.

HORIZONTAL AND VERTICAL TRANSFER

Transfer of a lethal dose can occur between insects. This is referred to as *horizontal transfer*, and it occurs in social insects, such as termites, wasps, and ants, and non-social species, such as cockroaches. The actual transfer may be through mutual grooming; sharing of food; or eating of dead bodies, cast skins, or regurgitated liquid. Formulations with low-dose or slow-acting insecticides provide for the donor individuals to return to the nest site or harborage before the insecticide takes effect. *Vertical transfer* is between different generations of the same species, such as from adults to immatures.

Termites. The active ingredients in non-repellent termiticides and termite-baiting systems vary in their modes of action and how much they are transferred. Some of these insecticides are horizontally transferred from initially exposed donors to others in the colony. In general, lethal transfers of termiticides occur when the donor is exposed to high doses of the insecticide for extended periods.

- *Liquid termiticides.* The efficacy of some non-repellent liquid termiticides depends on the behavior between colony members. Termites first acquire a dose on their cuticle by tunneling through treated soil, or in their foregut by ingesting the active ingredient after moving soil particles with their mouthparts. Exposed termites transfer insecticide to nestmates by grooming. Workers are groomed by

Sublethal exposure to repellent and non-repellent insecticides can result in reduced tunneling, movement, and feeding in some termites.

their nestmates and, in the process, ingest the insecticide. These contaminated recipients may become secondary donors through horizontal transfer to other unexposed termites.

- *Baits.* The strategy of using baits for termite control is based on individual nestmates sharing food by regurgitation and defecation. To transport food from a foraging site to nestmates, there is oral-to-oral and anal-to-oral transfer of gut fluids. Termites that consume insecticides in a bait matrix will spread it by food sharing with others in the colony.

For most termiticide baits, transfer occurs within the first 24 hours, and there is usually secondary transfer to other nestmates.

- *Microcapsules.* The grooming behavior between nestmates can improve the efficacy of encapsulated termiticides. Microcapsules on the cuticle of worker and soldier termites are removed by the grooming of nestmates and are broken in the process. Some of the microcapsules removed from the antennae and forelegs are broken during the process of self grooming, which usually involves the mouthparts.

Cockroaches. German cockroaches redistribute ingested insecticide from toxic baits within the harborage. Foraging adults transfer the active ingredient of baits to the harborages where nymphs are aggregated. First- and second-stage nymphs are killed by consuming contaminated adult feces. This transmission primarily targets small nymphs, because they remain in or near harborages and feed on feces.

Pheromones excreted in the feces that sustain aggregations also play a role in the transfer process. Early stage nymphs are the most responsive to the aggregation pheromone produced by females.

Ants. Worker ants gather and redistribute liquid carbohydrates and proteins gathered during foraging. Food sharing generally occurs for all foods, but especially carbohydrates. This may be because of the high energy requirements of foraging workers, or the demands of the developing larvae. Liquid ant baits used in household pest control may have only carbohydrate ingredients or both carbohydrate and protein components.

Granular bait is also shared among workers. Workers of the same colony may not transfer solid food, but they can exchange liquid extracted from the bait. In some species, foragers give food items to an intermediate nestmate at the nest periphery, rarely bringing a food into the nest themselves.

Pharaoh ants rarely exchange food outside the nest.

Ant food sharing

SECONDARY TRANSFER

The attachment of insecticide residue to the insect is not always strong, and a small amount can be dislodged during normal activity. This may result in secondary deposits in harborages or aggregation sites. Dislodged material is available for pick up by untreated individuals, but secondary exposure may not be lethal because the amount may be too small.

Transfer of particulate active-ingredient material may occur by direct physical contact between treated and untreated individuals. Typically this is from WP and SC formulation residues on nonporous surfaces and the initial pick up of a large dose. This type of secondary transfer is most effective against ants and cockroaches that are often in crowded harborages for long periods. This type of secondary transfer is dependent on individual pests acquiring a sufficient dose of insecticidal powder or crystal to provide for dislodgeable residues.

Argentine ant workers will exchange food and granular bait between 4 to 12 other workers; for this species the granule size must be small to be accepted.

CHAPTER 8

APPLICATION ECONOMICS

In professional pest control, emphasis is usually placed on the technology of killing pests, while the cost, or economics, of the process is often overlooked. However, designed into every application tool and insecticide is the concept of *application economics*. This concept considers how much insecticide is being applied, and how well the application tool works in terms of amount and cost. Application technology covers the control effectiveness of treatment, while application economics covers the cost effectiveness of it—both are important.

The "more is better" approach to pest control must be replaced with the "less is more" approach. By the time a professional service provider is called, pest problems have usually become severe, and customer attempts failed. While professionals have the advantages of training, equipment, and application, the customer often expects "stronger" insecticides and more treatment. The fact that precise or minimum application will achieve control is often difficult to understand for both technician and customer. However, applying more does not provide better control; more is not better, it is certainly wasteful and usually expensive.

Disregard for the repair and maintenance of application equipment does not always result in system failure or chronic breakdown. But it can have subtle consequences, such as a change in flow rate, delivery pattern of a nozzle, pressure maintained in a tank sprayer, or a valve that leaks due to worn washers. These performance changes are usually slight and go unnoticed or disregarded. However, they should be treated as more than simply an inconvenience, because they can cause a costly waste of material and result in misapplication or non-target exposure.

OVER APPLICATION

Over application occurs when more insecticide is applied than is necessary to achieve pest prevention or control. It can be a service technician spraying to run-off as liquid insecticide drips from a surface, dusting to produce a thick visible layer on the surface, re-tracing a spray pattern and over-applying to carpeting, or unnecessarily treating a section of baseboard or wall void. It may be only a few additional ounces of liquid or dust during a service call, but when it is routine with every service it becomes an expensive habit.

Insecticides are mixed by the gallon and wasted by the ounce.

Over application of insecticides is a common and unrecognized practice in professional pest control. It is difficult to see because it is only a small amount, perhaps just ounces at a time. Modern application tools are lightweight, ergonomically designed, and easy to use. However, a comfortable grip and a responsive valve trigger can be a problem as much as a benefit. Sometimes ease of squeezing a comfortable trigger can result in over application and extra cost.

Example: A technician services 15 accounts a day, and at each account applies insecticide for an extra 30 seconds with a coarse fan spray. In a single day, a gallon of insecticide is over applied, which is 20 gallons a month.

Liquid Insecticides. Spraying insecticide into or adjacent to pest harborages is a standard pest control practice. The application equipment combines pressurized tanks, valves, and nozzles to deliver a measured amount. However, there can be considerable variability in application, due to the habits of the technician and the type and condition of the equipment.

- *Surface application.* The objective of surface spraying is to establish a residue. Treating surfaces to the point of run-off is unnecessary; it wastes insecticide and increases costs. Control is achieved when a lethal amount is picked up on the feet of crawling insects, or a lethal concentration penetrates a porous surface, such as wood, to intercept larvae of wood-infesting beetles. A spray that produces a pattern of droplets on the surface can be effective and efficient. A crawling insect walking across the surface picks up a lethal dose of insecticide in a short distance of travel or short time on the treated surface.

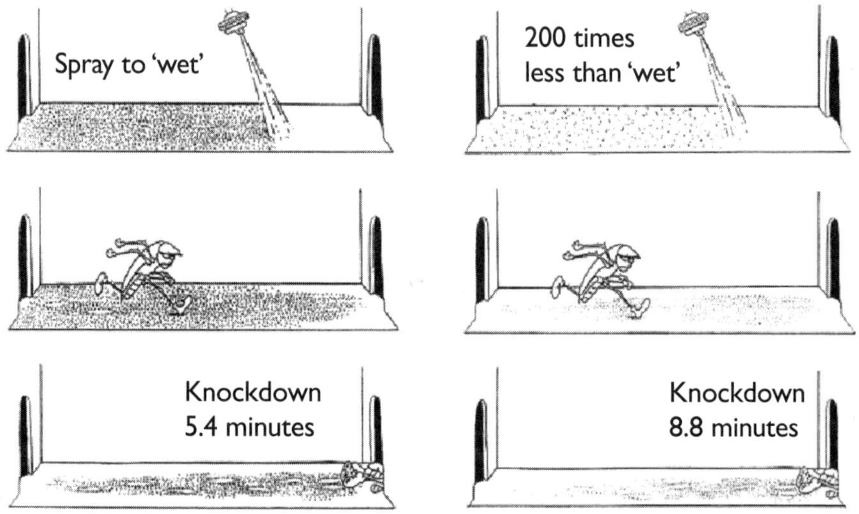

Example: After a surface is sprayed to the point of being "wet," German cockroaches walking on the dry residue are knocked down in 5.4 minutes. When the amount of insecticide applied is 200 times less than wet, the cockroaches are knocked down in 8.8 minutes. By applying 200 times less insecticide, the time to knockdown and kill increases by only 3 minutes, while the amount of insecticide used and cost is considerably less.

- *The coarse and fine fan sprays* on standard nozzles are designed to deliver droplets to the treated surface. The pattern of dry droplet residue is sufficient for the transfer of a toxic dose of insecticide to the feet of any insect that walks on it. Within a short walking distance, a lethal dose is acquired. Spraying more does not increase efficacy, but does increase costs.

- *Crack-and-crevice application (C+C).* This is the most effective and efficient method for treating pest harborages. It treats only the surfaces regularly contacted by the pest, increases residual activity of the insecticide, and decreases exposure to people and pets. A C+C tip can be inserted into narrow openings to deliver a small amount of insecticide directly to concealed surfaces. The alternative treatment method is to use a pin-stream on the nozzle. This is less efficient and more expensive than a crack-and-crevice straw, as it is more likely to result in over application.

  Using a pin-stream for a C+C application is not cost effective because the flow rate (8.9 ounces per minute) of the pin-stream orifice is significantly higher than the crack-and-crevice straw (2 ounces per minute).

Example: Using a tank sprayer and C+C straw, a technician can treat 20 linear feet of baseboard in about 1 minute. About 2 ounces of insecticide are delivered while slowly pulling the C+C straw along the inner back of the baseboard. Using a pin stream to treat the baseboard, nearly 9 ounces of liquid would be applied, and some would splash on the wall during application. The extra 7 ounces is wasted and costly.

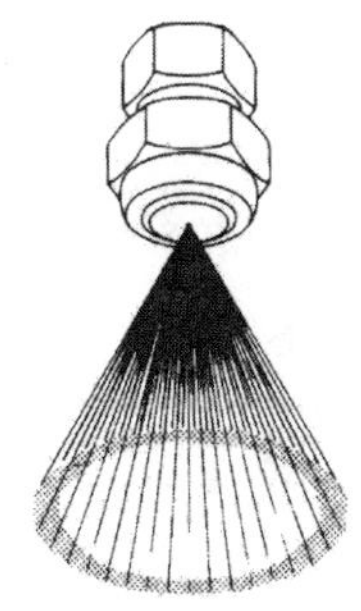

Cone Spray

- *The objective of treating behind baseboards* is to create a continuous band of insecticide residue on the void surfaces. This can be done by pulling the C+C straw along the inside of the void while spraying. Treating at intervals will leave untreated gaps on the surface and provide safe resting sites for pests.

- *Perimeter application.* Treatments of foundation walls and perimeter soil are typically done with a backpack sprayer because of the capacity and carrying convenience. Insecticide labels may recommend spraying 1 foot up the foundation wall and 1 foot out onto the soil. This can be done with a coarse fan spray and holding the nozzle 14 to 18 inches from the wall. However, a common practice is to use a pin-to-cone nozzle. These were not designed for treating flat surfaces and have a high flow rate, which can result in over application.

  The objective of perimeter treatments is to establish a barrier of insecticide residue on surfaces to prevent or reduce entry of crawling pests. A fan-spray provides a linear pattern of droplets without the circular overspray of a cone nozzle and the potential of drift to non-target surfaces.

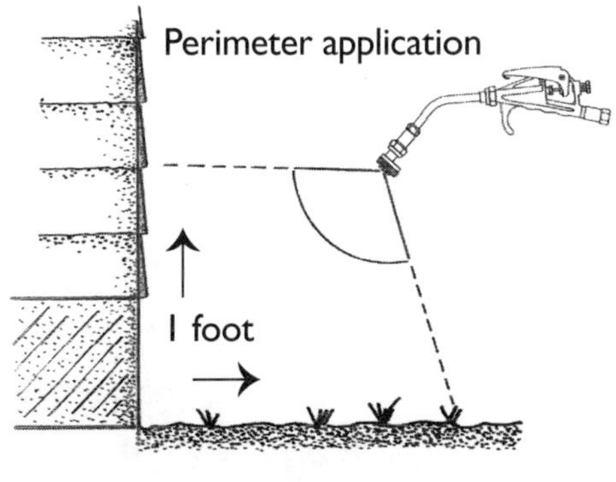

Example: Label directions specify 1 gallon per 1,000 linear feet for perimeter application. Treatment can be done with a coarse fan spray (17 ounces per minute) and a 2 mph walking pace. A cone-spray nozzle can deliver 45 ounces per minute, which is more than 2.5 times the coarse fan delivery rate. Using the cone spray results in costly over application and does not give the 1-foot-up, 1-foot-out spray pattern.

- *Termite control.* Over application of liquid termiticide occurs when applicators are careless with the amount applied during each soil rod insertion, or when applicators use tools without trigger- or lever-action valves and directional nozzles.

  Soil rodding with a nozzle that has a flow rate of 2 gallons per minute requires a treatment time of 12 seconds for a 1-foot deep foundation, and 48 seconds for a 4-foot deep foundation. The lever-action on/off control on professional valves provides precision application. Soil rods with a ball-valve control are less responsive, requiring 1 to 2 seconds for the applicator to close the valve. This extra time may seem small, but can result in costly over application.

Example: Treatment of a house with 160 linear feet and a 4-foot deep foundation requires 48 seconds of liquid application for each soil rod placement. This would deliver the label rate of 4 gallons per 10 linear feet, or a total of 256 gallons. Taking an extra 2 seconds to close a ball-valve would deliver 266 gallons, or an expensive over application of 10 gallons.

Example: Over application can occur when applicators fail to monitor the amount of treatment time during soil rodding. An average of 1.5 seconds of extra time with each rod insertion when treating a house with 160 linear feet will result in an additional 8 gallons of termiticide applied.

The efficacy of modern termiticides depends on establishing a continuous (no untreated gaps) and uniform (concentration) zone of treated soil around a structure. This zone can be established by using tools and nozzles that compensate for various soil conditions and reduce unnecessary and costly over application.

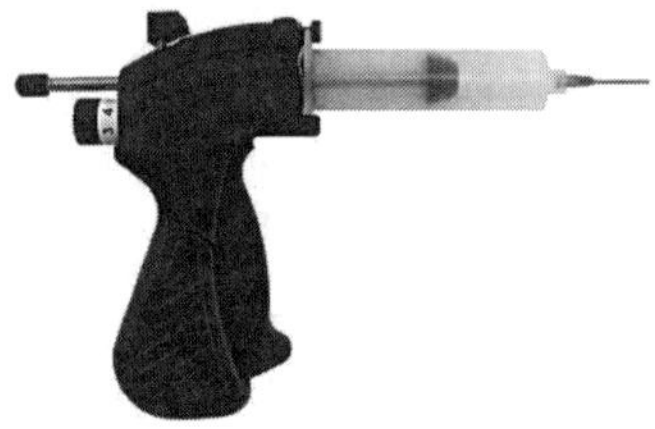

Bait gun

Baits. Gel baits are an important component of ant and cockroach control programs, but their efficacy depends on consistent application. Using modern baits requires consistent delivery of small amounts, 0.25 to 0.5 grams, to pest harborages. Precise application increases control, eliminates waste, and reduces overall operator costs. However, in spite of these benefits, operators may use bait guns that deliver variable size placements, instead of guns that are adjustable to ensure delivery of the same size placement with each pull of the trigger.

Example: A technician has a 30-gram syringe of cockroach bait and a variable-delivery bait gun. Four accounts must be treated, each with 15 bait placements. One syringe will do all four using the label rate of 0.5 grams per placement. But with a little more trigger squeeze, placements can be 0.7 grams. Instead of 7.5 grams total bait for the first account, 10.5 grams is placed—an extra 3 grams. The second account gets the same treatment, again with 3 grams extra. At the third account, the syringe contains bait for only 13 placements, and no bait remains for the fourth account. Applying extra bait in each placement reduced the 60 placements per syringe to 43 placements.

Precision-application bait guns deliver a consistent amount of bait with each trigger pull. This enables technicians to treat inside infested harborages and achieve better control. It is not necessary to visually monitor every placement to ensure the correct amount. The economic benefits of using precision bait guns include: less time spent on application, less wasted bait due to over application, and more effective control.

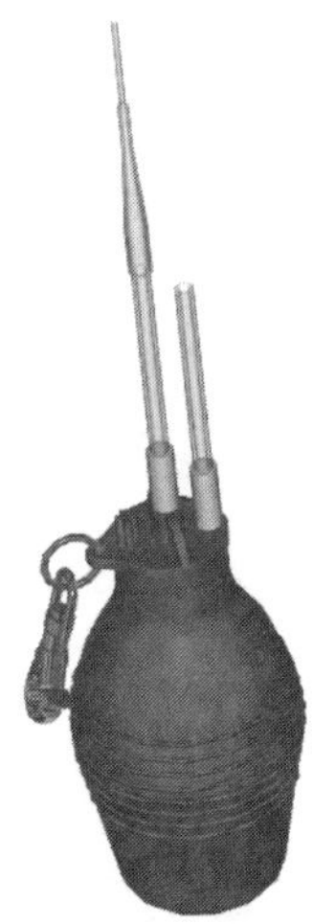

Duster

Dust. Over application of dust is common in professional pest control. Applying dust is like applying liquid: a thin layer of dust on the surface provides the most effective treatment. The physical nature of dust and the concentration of active ingredients are sufficient for a crawling pest to pick up a lethal dose with only limited contact. But dust is often difficult to see on treated surfaces, which can lead to the technician applying more. Over application can be unsightly and result in exposure to people and pets, but as important, it reduces efficacy.

Dust application is made unnecessarily difficult because of product label directions. The application rates are often presented in a format that promotes over application. Dust is delivered primarily as a crack-and-crevice application. However, on product labels the application rates are typically given in grams per square yard or pounds per 1,000 square feet. This is an inappropriate scale for indoor application.

The practical solution to the confusion created by product labels is to convert square yards and square feet to linear feet, and to use a hand duster with a consistent delivery.

Example: A dust product label recommends 2.0 to 3.0 grams per 1.0 square yard, which can be converted to 2.0 to 3.0 grams per 4.3 linear feet. The B&G Bulb Dust-R delivers 1.0 gram per squeeze. To apply dust using this duster, one squeeze per linear foot will deliver the high rate (3.0 grams), and one squeeze per 2.0 linear feet will deliver the low rate (2.0 grams) on the product label.

Example: A dust product label recommends 0.5-1.0 pound per 1,000 square feet, which can be converted to 8 to 16 ounces per 476 linear feet. This dust can be applied with the duster by using 1 squeeze per linear foot for the high rate (1 pound) and 1 squeeze per 2 linear feet for the low rate (0.5 pound).

To be effective and control cost, dust application has to be consistent from one technician to the next. The use of bulb dusters is subject to the habits of individuals and features of the unit, such as operational capacity and flexibility of the duster. Overfilling the bulb and excessive hand pressure can lead to costly and ineffective application.

Granules. The objective of perimeter treatment is continuous coverage of granules in a zone around and adjacent to the building. Achieving this without under or over applying depends on the equipment used. Handheld fertilizer spreaders can create a wide, broadcast-type pattern, easily applying twice the label rate. A piston-type applicator is better suited for perimeter treatment, as the delivery pattern is about 1.5 feet wide by 5 feet long. Because of this narrow treatment zone, the granule distribution is relatively uniform, which increases soil coverage and the potential for encountering target pests and decreases the potential for costly over application.

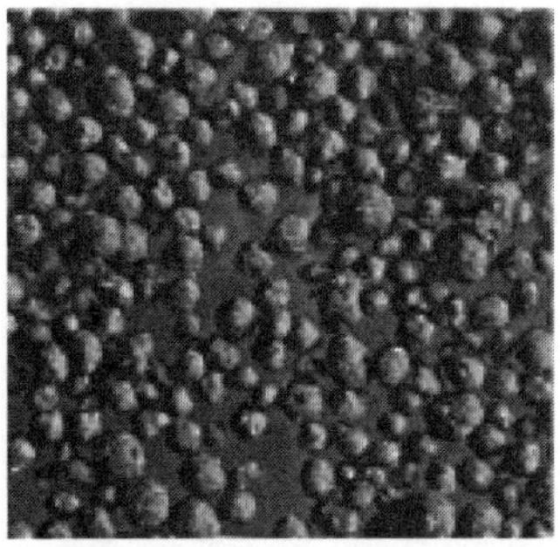

Effective control with granules depends on the size uniformity in the formulation. A wide variation in granule size can result in uneven application, because only small particles will pass into the siphon tube.

EXCESS APPLICATION

Excess application is unintentional and often occurs without being noticed by the applicator. It can happen when a piece of equipment is worn or damaged, with the result that more insecticide than expected is delivered during routine application. The excess amount may be small or the spray pattern only slightly altered so that the change is unnoticed. But with continued use, wasted ounces can accumulate to significant costs. Eventually the cost of the quantity of insecticide wasted during normal application exceeds the cost of replacing the nozzle or repairing the piece of equipment.

The use of old and worn nozzles is common in professional pest control. This practice results in gallons of wasted insecticide and increased costs. Technicians may periodically clean spray tanks and filters, but often fail to replace the nozzle on a regular basis. A 49% to 63% increase in flow rate can occur in a brass nozzle after 300 hours of normal use. However, even a slight increase in flow rate can result in excess application and an increase in the cost of treatment.

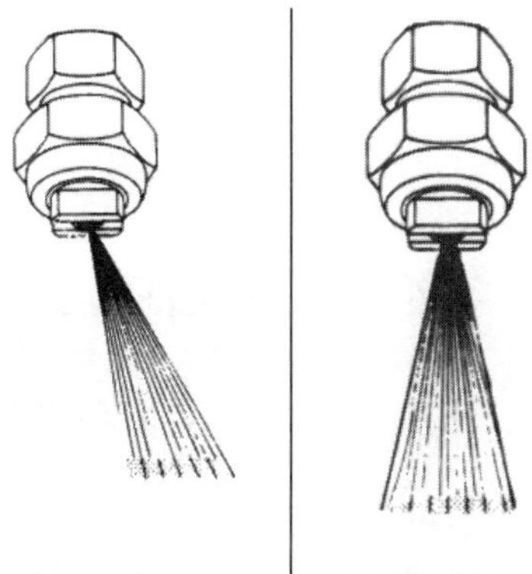

One-sided | Narrow width

Nozzle Wear. The four-way nozzle commonly used on pest control sprayers can show some characteristic conditions of wear with extended use. The most common are the hot spot, or narrow width, and one-sided.

- ***Hot spot/narrow width.*** This condition is characterized by one or more locations in the spray pattern receiving a large quantity of liquid, and the pattern width reduced to 12 inches (from the normal 17 inches).

- ***One-sided.*** This is characterized by a large quantity of spray liquid shifted to one side of the pattern. The width of the spray may remain close to normal, so this may not be apparent when spraying.

Nozzle Flow Rate. When a coarse fan nozzle has a 12% change in flow rate, the amount of liquid delivered increases from 17 to 19 ounces per minute. While this may seem to be a small amount, the excess application can be costly over time. Commonly encountered flow-rate changes in nozzles range from 12% to14%.

Example: Servicing 12 accounts a day, a technician uses a fan-spray nozzle to treat for 3 minutes in each account. The amount applied each day with a new nozzle is 612 ounces. Using a worn nozzle with a 12% increase in flow rate, the amount increases to 684 ounces. This is an excess of 72 ounces every day and nearly 3 gallons in five days.

A nozzle will show little visible indication of wear during its term of use, but deterioration of the orifice can be expected. Replacing the nozzle on a regular schedule will eliminate guessing and prevent waste.

Termite Control. The 180° and 360° directional tips on soil treatment rods wear just like any other nozzle. Modern termiticides can contribute to nozzle wear and increased flow rate. Regular replacement of these nozzles will prevent excess and costly application.

Example: A 160-linear-foot house with a 4-foot-deep foundation is treated with liquid termiticide. At the label rate of 4 gallons per 10 linear feet per foot of depth, it requires 256 gallons. If the 2-gallon-per-minute tip (4.26 ounces per second) has 10% wear, an excess of 25 gallons will be applied to the structure.

The combination of repeated insertion in clay and/or sandy soil, and the flow associated with the large volume of liquid in a termiticide application can accelerate nozzle wear in termite-control tools.

Excess dust application will not increase efficacy or residual activity, but it is likely to result in deterring the pest from the surface and reducing efficacy.

RESIDUE EFFICACY AND RETURN ON APPLICATION (ROA)

The objective of most applications is to establish an insecticide residue on surfaces that will be contacted by the target pest. The long-term efficacy of residues is an important aspect of control because the activity and movement of the pest is unpredictable. This "residue" feature of liquids, dusts, and bait formulations extends the initial application for days, weeks, and years in the case of termiticides. However, the full extent of residue efficacy depends on proper application, environmental conditions, and physical features of the treated surface.

The economics of insecticide application considers the time spent on application and the resulting residue efficacy time. This efficacy time is the *return on application* (ROA): the amount of residue time gained from the original application time. In the ROA, application time is invested to gain lethal residue time, and the goal is to have a long residue time relative to the time spent on application. The longer the lethal residue time, the better the ROA.

Liquid Insecticides. The longest ROA is typically provided by wettable powders (WP), followed by suspension concentrates (SC), then microencapsulated (ME), then emulsifiable concentrates (EC).

WP and SC formulations leave particles of the active ingredient exposed on the surface. These are easily transferred to the legs and feet of the pest and deliver a lethal dose. Capsules of ME formulations remain for an extended period until walked on and crushed by the pest, or they leak their enclosed liquid insecticide onto the pest or substrate. EC formulations generally establish a smooth surface residue that is contacted by the pest.

- *Surfaces features.* The uneven, pitted, and ridged surface of bare wood, latex-painted surfaces, brick, and concrete reduce the residual life of EC formulations. The amount of surface residue of wood-protection insecticides (typically EC) significantly decreases on wood treated in crawlspaces and attics. The oxide layer on metals, including stainless steel, reduce the long-term efficacy of most insecticide formulations, but less so for WP and ME.

Dusts. Insecticidal dusts have the longest ROA. Undisturbed residues typically maintain their efficacy for months. The active ingredient is relatively loose on the surface and is easily transferred to pests as they walk on the surface or groom their legs and antennae that have contacted a treated surface. Dusts can be fast acting because the insecticide-based products are taken into the mouth during grooming. Loss of body water and desiccation caused by silica-based dusts may take several days; the addition of pyrethrins for knockdown alleviates delay.

- *Surface features.* Dusts are generally unaffected by physical features of the treated surface, which is useful since the majority of dust-treated surfaces and voids are not seen during application. Wet surfaces, before or after application, can reduce the long-term efficacy of dust formulations. Because of this, modern insecticide dusts are formulated to resist wetting and retain their efficacy.

A common misconception is that bait placements are contaminated and become repellent when accidentally exposed to liquid or dust. This is not the case. Cockroaches will continue to visit and feed on bait placements that have been contacted by insecticide droplets or dust particles.

Gel Baits. Baits can have a long ROA, but the objective of baits is to be eaten in 1 to 2 days. Typically, the pest becomes inactive and dies within 3 to 4 days. The ROA is influenced by the unique feature of gel baits: they have high moisture content at time of application, but about half of it is lost within 48 hours. While this moisture loss can make placements less appealing to foraging pests, it essentially doubles the concentration of the active ingredient in the placement. Cockroach adults and even small nymphs will feed on even months-old bait placements.

- *Surface features.* The residual activity of baits is only slightly influenced by features of the treated surface. Hot surfaces above room temperature can liquefy the gel matrix, decrease the stability of the placement, and quickly increase the loss of moisture. Dry and porous surfaces, such as paper or cardboard, may also remove water from bait placements and decrease their attractiveness to the pest.

Sticky Traps. Although they are not usually considered to be control methods, modern sticky traps can be effective in eliminating small infestations and preventing entry of occasional invaders. In these roles, sticky traps have a long ROA of months. They are quickly and easily placed, exposed or concealed, and remain effective as long as the sticky surface remains at least 50% to 60% open. In addition, the substrate and environmental conditions have no influence, except that of excess moisture, and they provide capture-control of a broad spectrum of pests.

CHAPTER 9

STRUCTURAL WOOD AND URBAN SOIL

The wood in a building frame and the soil adjacent to a building are the most commonly treated substrates in professional pest control. Most of the information provided in training sessions and manuals pertains to pest biology and habits, application methods, and equipment. Little is provided on the basics of structural wood and urban soils. While extensive knowledge of these two is not necessary for service technicians, the structure and function of wood and soil can be helpful in the application of termiticides and wood-protection insecticides.

STRUCTURAL WOOD

Tree species can be classified into two groups: hardwoods, with broad leaves that are shed every fall; and softwoods, with long, needle-shaped leaves retained throughout the year. Softwood timber, such as pine, spruce, and fir, is the dominant material used in building construction. This is the wood most commonly treated with fungicides and insecticides for long-term protection. Hardwoods are used in the construction of most shipping pallets and crates, and for hardwood flooring and furniture.

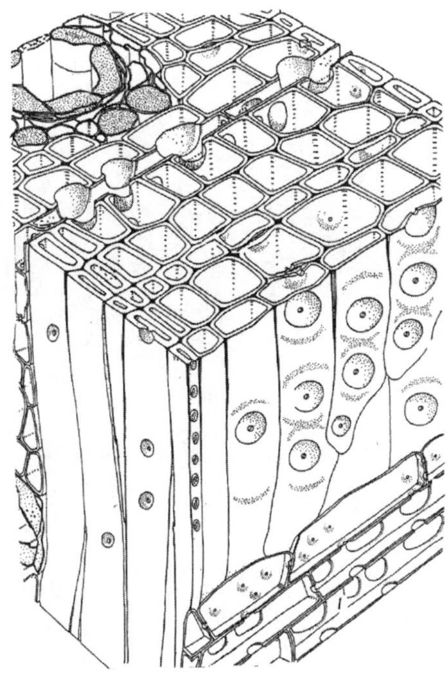

Softwood Lumber. A close-up examination of softwood lumber shows that it is composed of long, slender cells called tracheids. The tracheids are oriented parallel to the tree trunk and along the length of structural lumber. Tracheids are packed close together so that a cross-section of a piece of wood resembles a honeycomb. The role of tracheids is to conduct nutrients up the tree from the roots; this is done by the transfer of liquid between overlapping ends of tracheids.

Liquid insecticide can be injected into most softwood lumber, and it can be carried along the length of the piece by moving through empty tracheids.

In addition to tracheids, another type of cell is present in softwoods, the parenchyma. Carbohydrate food material is stored in the parenchyma cells. This material remains relatively stable in lumber for many years. In structural wood, such as pine, it is eaten by the larvae of wood-infesting beetles. When the wood moisture content is above 20%, this carbohydrate is used as a food by fungi.

Hardwood Lumber. Hardwood lumber is also composed of tracheids, vessels, and parenchyma cells that store carbohydrates. The vessels are large-diameter cells that are linked together longitudinally along the grain of the wood. The vessels remain open after the lumber is dried and are available for carrying liquids, such as insecticides and fungicides, injected into the wood.

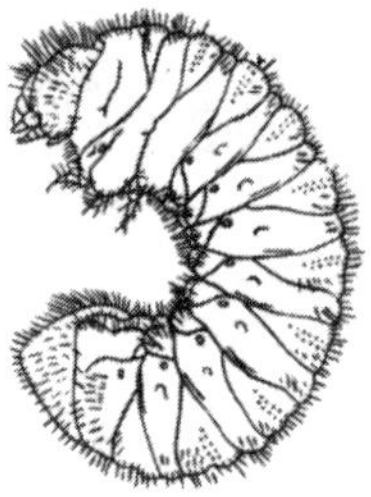

Powderpost beetle larva

Powderpost beetles infesting tropical hardwoods may complete development faster than those in hardwood from northern temperate regions.

More carbohydrate food is stored in the parenchyma cells of hardwoods than in softwoods. In tropical hardwood lumber, 50% of the wood volume is parenchyma, while hardwoods growing in northern regions may have only 24% parenchyma cells. The large amount of carbohydrates stored in hardwoods makes them an ideal food for the larvae of lyctid powderpost beetles.

Liquid insecticide injected into hardwood is easily carried along the grain of the wood by the large vessels and tracheids; this helps in creating a "chemical fence" to control or limit a powderpost beetle infestation.

MOISTURE CONTENT

The amount of water that is present in wood influences strength and stiffness, susceptibility to fungal attack, and the feeding of wood-infesting insects. The moisture content of wood in living trees and recently felled logs is about 40%. For wood used in building, the moisture content must be reduced to at least 12%. This level is suitable for construction and is below the 20% required for wood-decay fungi. However, 12% moisture is acceptable for wood-infesting beetles and termites that infest structural timber.

The moisture content of structural wood varies from 6% to 13%. The seasonal fluctuation of wood-moisture content influences the feeding and development of wood-infesting beetle larvae; they usually suspend feeding when wood moisture declines in winter.

LUMBER DRYING METHODS

General construction lumber is dried to a moisture content of 12% to 19%. Air- and kiln-drying processes are used to reduce the original 40% or greater moisture content of hardwood and softwood lumber to this level. Some drying kilns operate at temperatures up to about 212°F; high-temperature kilns operate above this temperature and are used to dry softwood lumber. Some construction lumber is air dried, especially in regions where the ambient humidity is low. Air drying is adequate for construction lumber if enough time is allowed to reduce the moisture content.

Liquid-insecticide application to the surface of structural wood penetrates to the region of concentration of carbohydrates; this helps to control the first-stage larvae that feed in this outer layer.

Softwood Lumber. The air- and kiln-drying processes for softwoods concentrate carbohydrates and other sugars at the surface of the lumber. This concentration provides an important food source for early stage larvae of wood-infesting beetles. It is the first layer of wood they feed on after they hatch from the egg and begin chewing into the wood surface. Old house borer and anobiid powderpost beetle larvae tunnel and feed in surface wood before going deeper.

Hardwood Lumber. Much of the starch and other carbohydrates in the wood that can serve as food for wood-infesting beetle larvae are eliminated by air drying. The slow air-drying process consumes these materials. As a result, this wood is less desirable for lyctid powderpost beetles and less likely to be infested. The female beetle typically bites and chews a small sample of the wood before laying her eggs. In the case of air-dried hardwood, she may determine the starch content to be too low for the development of larvae and not lay eggs on exposed surfaces.

In kiln drying, there is little or no loss of natural starch in the wood. Hardwoods that are kiln dried, which is a common process for most commercial hardwoods, remain susceptible to attack from lyctid powderpost beetles for many years.

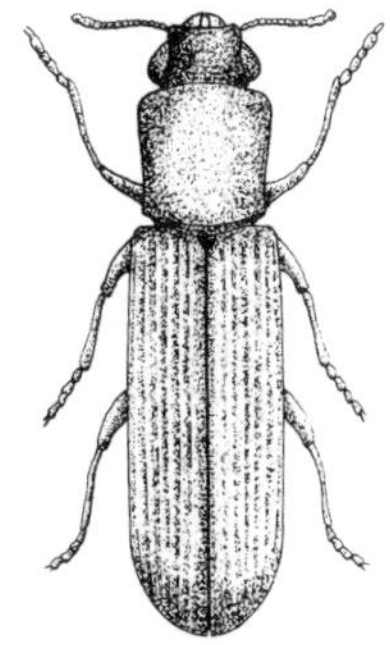

Lyctid powderpost beetle

LIQUID PENETRATION

Injection. Liquids can be injected below the surface of seasoned softwoods and hardwoods. The objective is to force the liquid into and through the open tracheids. Variations in their diameter and the connections between tracheids interrupt the flow of liquid. There is "critical distance" of about 10 to 20 inches when all the continuous flow pathways of the large tracheids have been followed, and there is little additional movement of liquid.

Structural lumber infested with termites, powderpost beetles, or carpenter ants can be protected from further damage by the injection of liquid insecticide into wood at the edges of the active colony galleries or larval-feeding tunnels. This treatment creates a "chemical fence" in the uninfested wood.

It is a common misconception that liquid must be injected into wood under high pressure of 60 psi to achieve movement along the wood grain. Suitable flow can be obtained with the 30- to 35-psi pressure of a standard tank sprayer.

Surface Penetration. Penetration of liquid into wood following application to the surface is based on diffusion. This involves movement from regions of high concentration to regions of lower concentration. The rate is directly proportional to the concentration gradient. The more insecticide that is applied to the surface the more penetration there will be below the surface.

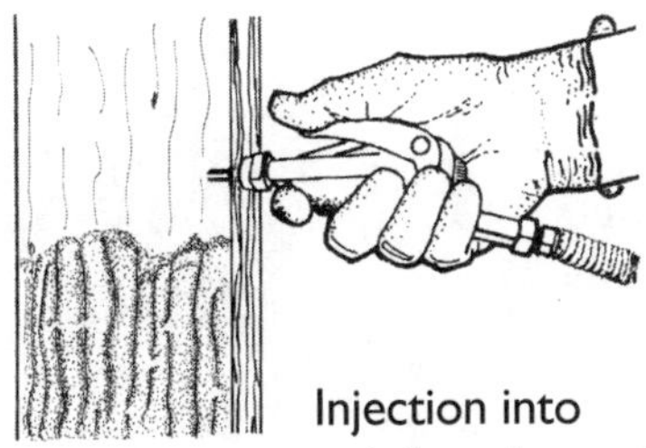

Injection into uninfested wood

Immersion. The dip-diffusion process is usually a short-time immersion followed by drying to allow the preservative to diffuse into green timber. High concentrations are used so that the surface of the wood holds a reservoir of chemical, which allows penetration to a satisfactory depth. Prolonged immersion methods typically use low concentrations of preservative.

WOOD PROTECTION CHEMICALS AND METHODS

Wood preservatives can be categorized into organic or solvent-based compounds, and inorganic or water-based compounds. The insect-toxicity mechanisms of these materials involve denaturing of proteins, inactivation of enzymes and proteins, and disruption of cell membranes. The degree of protection provided by a preservative depends on penetration and permanence in the wood. The wood species and application method, such as surface treatment or pressure treatment, influence the final penetration.

Solvent-Based Preservatives

- *Creosote.* This preservative is derived from coal tar and packaged as an oily liquid with a pungent odor. There are grades of creosote based on their viscosity and volatile components. Type 3 has low viscosity and is suitable for surface application. The light oil qualities of this material allow for improved penetration into wood. It is applied to fence posts, landscape timbers, and utility poles. Creosote-treated wood has a service life of 30 years.

- *Pentachlorophenol (PCP).* This is an oil-based wood preservative. It is widely used for treating utility poles. When formulated as a mineral oil, the chemical migrates to the wood surface and may form crystals there. The formation of PCP crystals on the surface can be an irritant to workers exposed to treated wood. The availability of alternative products and environmental concerns has resulted in limited or excluded use of this preservative.

- *Copper 8-quinolinolate.* This solvent-based preservative is used as an industrial wood preservative to control sap stain and to protect wood against mold and mildew. It can also protect against insect damage for wood used above ground. Copper 8-quinolinolate is often used as a wood-surface coating. It is impregnated into wood by both non-pressure and pressure techniques.

Water-Based Preservatives

- *Chromated copper arsenates (CCA).* These preservatives have replaced the oil-based materials such as PCP and creosote. Chromated copper arsenates, ammoniacal copper arsenate, and ammoniacal zinc arsenate are commonly used to treat dimension lumber for outdoor structures, such as decks and playground equipment.

- *Alkaline or ammoniacal copper quaternary (ACQ).* This preservative is made up of copper, a fungicide, and a quaternary ammonium compound. It has come into wide use following restrictions on CCA. National and international standards determine the volume of preservative uptake required for a specific timber-end use.

- *Copper Azole (CA-B, CA-C).* This preservative is similar to ACQ, the difference being that the dissolved copper preservative is augmented by an Azole co-biocide instead of the quat biocide used in ACQ. The Azole co-biocide yields a Copper Azole product that is effective at much lower retentions than that required for equivalent ACQ performance.

- *Borates.* Boric acid and borax prepared as a 46:54 mixture (disodium octaborate tetrahydrate) is the most common form for borate wood preservation. Other forms of boron preservatives include those soluble in organic solvents, including tri-hexylene glycol borate and borates prepared in mono-ethylene glycol. Borate compounds can leach out of wood when immersed in water for long periods. However, once below the surface, they remain stable and are less prone to leaching.

URBAN SOIL

The elevation, grading, and plant cover of the layer of topsoil around a structure influence the distribution of pests and pest control. Plant roots and the burrows of small vertebrates and invertebrates penetrate this layer of soil that was applied after construction. The conditions of this soil are determined by rainfall and sunlight, and both can have critical influence on insecticide application and residual activity.

Termiticide application to the perimeter soil around buildings is influenced by soil structure, composition, and soil moisture. Product labels anticipate the conditions and permit changes to compensate for them. For example, clay soils can present a problem because they may not accept the application volume specified by the product label.

SOIL STRUCTURE

Soil Types. The proportions of sand, silt, and clay in soil are used to define soil types. In typical soils the sand grains are coated with a film of microscopic clay particles, and the clay forms connections between sand grains and silt. Organic matter and humus play a major role in the structure of soils. Without humus, soils with a high percentage of clay and silt become very dense. Humus gives soil the dark color that is usually associated with high fertility.

- *Sand.* Sand forms the framework of soil and gives it stability when in a mixture with silt and clay. Sand particles range in diameter from very fine, 0.05, to very coarse, 2.0 mm. Soil textures defined by the percentage of sand include loamy sand, 70% to 85%; sandy loam, 50% to 70%; and silt loam, 20% to 50%.

- *Silt.* Silt particles are too small to be seen with the unaided eye. The presence of silt gives soil a smooth texture without being sticky. It is easily suspended and carried in flowing water, but settles when the flow is reduced.

- *Clay.* This component of soil is chemically and structurally different from sand and silt. Clay has a large surface area due to the flat, plate-like shape of the individual clay particles. Each clay particle is actually a stack of microscopic sheets. Clay soils are dense and have small pore spaces.

A common misconception is that soil color is related to texture, such as the phrase "red clay." However, reddish brown soil may not indicate that it is a clay soil; very sandy soils may have a red color.

Soil Movement. Swings between hot, dry weather and severe rain or snow can affect the soil beneath buildings. Clay soils shrink during droughts and swell during floods, causing structures to shift. The adhesive properties of sandy soil decrease in dry conditions and pull away from foundations. Such sinking, or subsidence, occurs gradually, but swings from very wet to extremely dry weather accelerate it.

Signs of sinking soils are: problems in closing doors and windows, chimneys, or porches separating from the house, and bowing of basement walls.

SOIL TEMPERATURE

Soil temperature influences plant growth and the presence of insects and other arthropods. Many pest arthropods spend part or all of their lives in soil. Some feed on the roots of plants or decaying organic matter around the root system; many are scavengers or predators of other animals. Ants, solitary bees, and subterranean termites use soil as nest sites and foraging areas. Soil temperature can also influence the residual activity of pesticides applied to the surface.

A common misconception in the use of mulch is that it preserves soil moisture. Organic mulch will reduce evaporation loss in the first days or weeks following placement, but over a long dry period, the water loss from mulch-covered soil approaches the same amount as that from uncovered soil.

Winter Survival. For many insects, surviving winter depends on finding an insulated harborage. Snow acts as a good insulator, a reflector of solar heat during the day, and an excellent radiator at night. For example, air temperature above snow on clear nights may be below freezing, but the ground beneath remains at or slightly below the freezing point of water. In sandy soil the majority of over-wintering insects hibernate at a depth of 12 to 24 inches, but in clay soils they usually hibernate in the top 8 inches.

Ground Cover. Organic mulches can influence soil temperatures by shading it from the sun. They tend to reduce extremes in soil temperatures, and may indirectly favor the survival of soil-inhabiting insects. In periods of hot weather, organic mulches generally keep the surface of the underlying soil cool. In fall and winter, mulch can modify rapid temperature declines.

Landscaping around the perimeter of buildings influences soil temperature and reduces water evaporation. This band of soil may have higher moisture content than soil farther from the foundation. Wet soils have a higher capacity to retain heat and are slow to change temperature. These soils are favorable for ants, termites, and other soil-nesting insects.

Freezing. Permeability to water is almost stopped when saturated soils are frozen, however soils frozen after drainage or frozen in a loose condition are permeable to water. In addition, soils with high organic matter are porous, and easily penetrated by water, even when the soil is frozen. Soil around the perimeter of buildings may be artificially stratified. The upper layers may have high percentages of organic matter, enabling them to retain permeability, but lower layers may be significantly less porous.

Manufacturer label instructions for termiticides recommend that application is not made to frozen soil. The liquid termiticide may not be absorbed by the soil, resulting in run-off or an incomplete treatment zone.

LIQUID MOVEMENT

The rate of movement and the depth to which water moves in soil are influenced by the pore spaces, that is the spaces between the solid particles in soil that contain water and air. Pore spaces are large in soils composed of sand particles, and small in soils composed of clay. Fine soil has smaller pore spaces than coarse soil, but it has a greater number of pore spaces than coarse soil. Coarse sandy soil has large pores, enabling it to conduct water flow faster than fine-textured, clay soil.

Termiticide Application: Trenching. Not all termiticide applications require rodding of the perimeter soil. Some buildings have foundations that are at ground level or only slightly below grade. These shallow foundations can be treated by trenching and applying liquid termiticide to the trench. Label directions usually specify a 6-inch wide and deep trench. The application objective is to establish a continuous zone of treated soil around the building, next to but not below the footer.

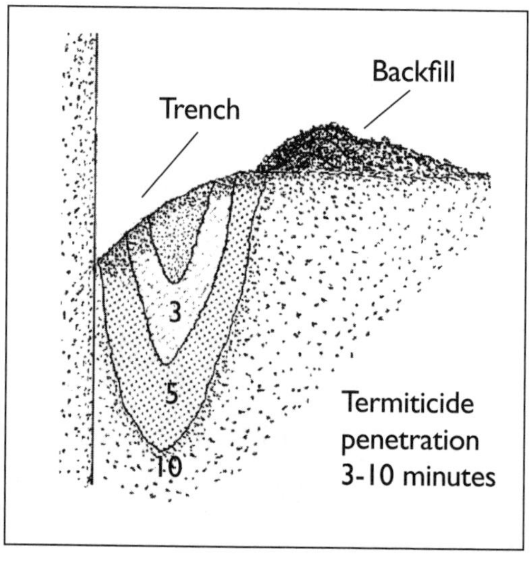

Liquid termiticide penetrates soil in a predictable pattern. Liquid moves slowly in the soil, from saturated zones to unsaturated zones, with the leading edge of the termiticide-saturated soil advancing into the dry soil surrounding it. Eventually the entire trench is saturated, from the point of application to the bottom edge of the soil in the trench. Slow application of liquid to the trench causes it to gradually advance to the edge of the pattern. This feature of liquid flow, from saturated to unsaturated regions, is especially important in treatment of clay soils.

Termiticide penetration in the trench soil can be slow, and it may take several minutes for the soil to accept the liquid. Flooding the trench at once with the full amount of termiticide may not be as effective as applying the full amount in two or three stages.

Termiticide Application: Soil Surface. Pre-construction treatment requires the establishment of a complete horizontally treated zone. This applies to all soil surfaces to be covered beneath a concrete slab. The surface may be the top of the original soil or a covering of coarse gravel. The soil type, such as loamy sand or clay loam, influences the vertical movement or soil penetration of the liquid termiticide.

The penetration time and depth depend on the soil texture. There can be a temporary inhibition to penetration when a layer of sand or clay is below the soil surface. Liquid flow is initially slow from a fine-textured layer to a coarse-textured layer. Water does not move easily from soil that has water-filled pores into a sandy soil that has large air-filled pores or to a dense clay soil.

Unsaturated Soil. Water flow in unsaturated soils is slow. In general, water flows readily from pores in wet soil zones to pores in dry soil zones. Water movement in unsaturated soil can be upward, downward, or horizontal, depending on the position of a dryer zone in relation to a wetter zone.

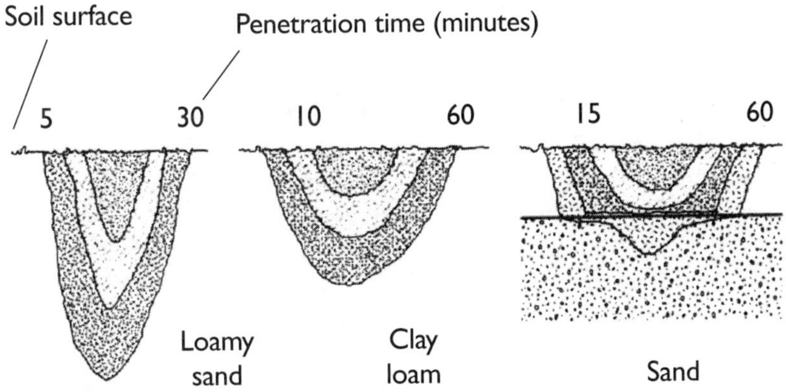

Soil penetration patterns

Saturated Soil. In saturated soils all pore spaces are filled with water, and water flow is generally rapid. When soil pores are continuously connected to each other, they form a natural conduit for water to move through the soil. Gravitational forces influence the flow of water in saturated soil, and the direction of flow is generally downward.

Termiticide labels do not permit application to soil that is water saturated.

Clay Soils. Clay is a term commonly used to describe a soil that is particularly resistant to water penetration. However, clay soils have the highest moisture-retaining capacity. Infiltration of water in clay soils is limited because most clay soils swell when exposed to water and this makes the pores smaller and reduces water movement. Dry clay soil will take up a large amount of water during the initial exposure, but it will take up less with continued exposure.

Water movement downward in clay soil may not occur until the upper layer of soil becomes saturated, then liquid will enter the lower layer. During the accumulation of liquid at the junction of the two layers, water may move laterally along the interface.

Following building construction, a layer of topsoil may be placed over a base of sandy or clay soil. Clay soil may restrict the flow of water from the surface layer. Liquid termiticide injected into sandy loam soil may move slowly or pool at the interface with the clay soil.

INSECTICIDES IN SOILS

The behavior of a liquid termiticide in soil plays an important role in determining its fate, persistence, and biological activity. The ability of a termiticide to provide termite prevention and control are linked to pH, moisture, leaching, and microbial degradation.

Soil pH. Most soils have a pH range of 4.5 to 8.0. The fate of insecticides in soil varies with pH differences, primarily because of the influence of pH on adsorption. Insecticide adsorption is usually higher in more acid soils than in alkaline soils.

Soil Moisture. In general, more insecticide is adsorbed when soils are dry than when they are moist or wet. The addition of water may cause the release of insecticides adsorbed to mineral or clay soils, but usually not in high organic soils. When water is added to an insecticide-soil complex, the water molecules begin to compete with the pesticide molecules for adsorption sites, and may force insecticide back into solution.

Leaching. The movement of a pesticide in soil with water is commonly known as leaching. This may occur in a downward, lateral, or even upward direction. The downward movement is the one of most concern for termiticide application. The important factors controlling leaching are water solubility of the insecticide, adsorption, soil type, and moisture. Insecticide solubility is one of the important aspects in the persistence of a termiticide in soil. Generally, insecticides highly or moderately soluble in water are leached from soils more quickly than those that are less soluble.

As water moves over sloping soil surfaces, pesticides previously applied to the soil surface can be desorbed and move with the run-off water. Lateral movement with run-off water is likely to be greatest in wet conditions when insecticides are least likely to be adsorbed.

Microbial Degradation. Soil bacteria are responsible for the decomposition of insecticides in soil. The microbial degradation process is influenced by the availability of moisture, nutrient level present in the soil, oxygenation of the soil, and adsorption of the insecticide to soil particles. The degree of chlorination of an insecticide may influence the biodegradability: the more chlorine in the molecule the more persistent. Most termiticides show a considerable period of time between application and the first detectable aspect of breakdown.

EXTERIOR AND INTERIOR FEATURES

The influence that exterior building features have on the presence and control of structural pests and perimeter pests is often overlooked. The soil surface and subsurface drainage can have a direct impact on the application of liquid termiticide to perimeter soil; soil settlement provides the sub-slab void for the application of liquid and foam; and the basic foundation system can determine the depth of soil rodding. Understanding and evaluating the conditions of the structure to be treated will make the application more effective.

Indoor environmental conditions influence the presence and long-term infestation of pests. Humid conditions are required for insects to survive and breed inside buildings, and excess humidity leads to insect and wood-decay problems. Carpet and fabric are the most common substrates in modern houses and office buildings. They provide harborage for dust mites, cat fleas, bed bugs, and dermestid beetles.

EXTERIOR FEATURES

Wood Frame Construction. Platform frame construction is the standard wood-frame building system used for residential buildings. A platform floor is built at each story, and load-bearing walls are erected upon it. Platform construction is typically made with nominal 2-inch, wood studs, and 2x4-inch timbers are commonly used for interior partition walls and exterior load-bearing walls. However, 2x6-inch timbers may be used for exterior walls to allow space for more insulation.

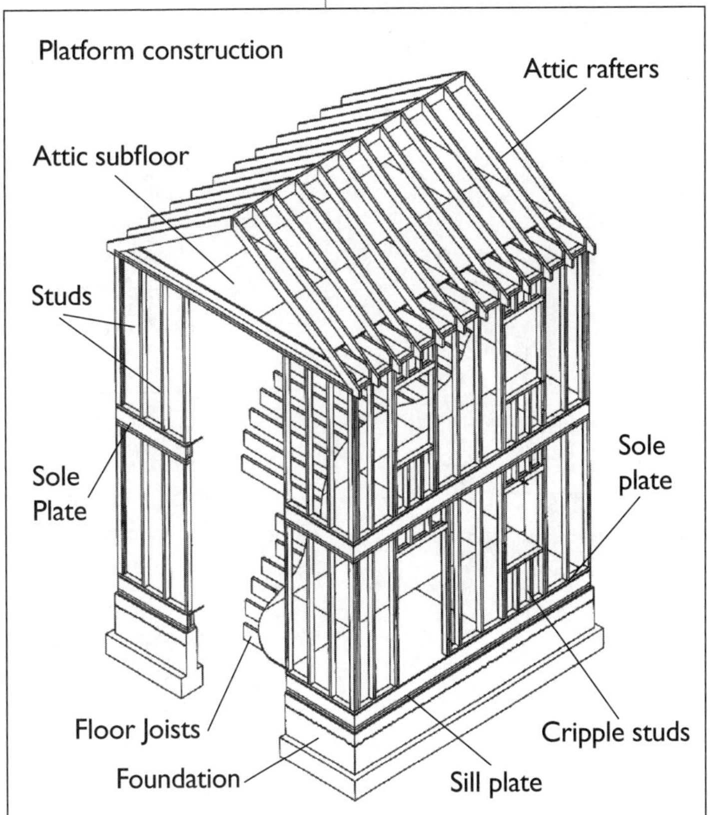

- *Joists.* These timbers are the primary structural members of a floor system. These are nominal 2x6, 2x8, 2x10, or 2x12 inch. Solid timbers, built-up beams, and glue laminated beams can also be used to form the basis of the floor support system.

- *Stud spacing.* A maximum 24-inch on-center spacing of wall studs is required by most building codes. Stud spacing of 16-inch centers is the most common, and should be anticipated unless otherwise known.

- *Sheathing.* This is a covering layer applied to the exterior wall studs to stabilize the structure. Sheathing is typically plywood or other manufacturer board. Sheathing provides a surface for the insulation and water vapor barriers, and for the exterior finish.

Soil Surface and Subsurface Drainage. The origin of water problems around buildings is often improper channeling of rainwater. The final slope of the construction site must provide for the movement of rainwater away from the foundation. Poor surface and subsurface drainage can lead to the containment of water around the house and

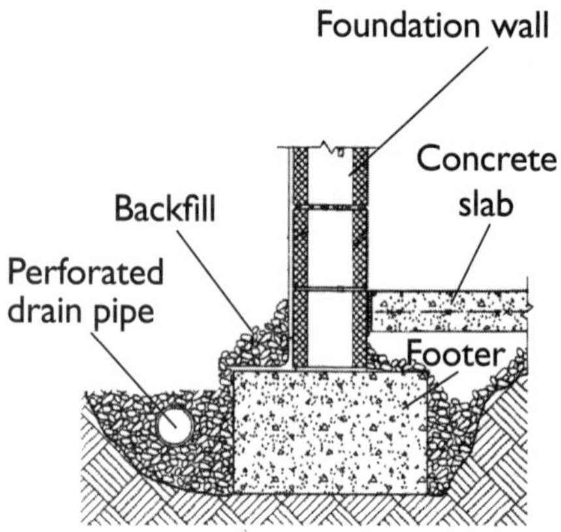

leaking of groundwater into the basement or crawlspace. This can result in structural damage to the foundation and contribute to termite infestations.

- *Foundation drainage.* Houses built in areas with poor-draining clay soils usually have a foundation drainage system. Drainage tile or pipes may be placed adjacent to the footing on a bed of gravel and covered with a filter fabric or a layer of coarse gravel. The cover prevents the drainage pipes from filling with silt. The outflow of foundation drainage is usually an open drainpipe, or it may flow into a gravel field or dry well.

- *Basement sump.* Drainage pipes inside foundation footers drain the water that seeps from beneath the foundation and through hollow block walls from the outside. Plastic or tile pipes or gravel-filled trenches may lead to a dry well outside the structure or to an interior sump pit, which is drained by a sump pump. Groundwater under the foundation and along the interior foundation walls is directed to the sump pit through gravity or pipes and pumped out.

Soil settlement beneath concrete slabs usually produces an irregular 1- to 3-inch void, which may extend under the entire slab. This space may have an uneven surface and include rocks, soil depressions, and utility pipes.

Foundation Systems and Features. Regardless of the soil type or water conditions, settlement of a building and surrounding soil is expected in most areas. The soil beneath the foundation carries the weight of the building structure, and the soil beneath the foundation floor carries the weight of the concrete.

- *Poured concrete.* These foundations have walls that are about 8 inches thick and 8 feet high. Walls are usually coated on the outside with waterproofing materials. Drainage tile or pipe may be installed adjacent to the footings.

Wood-sheeting boards, studs, and spacer blocks are used as forms for the concrete. These may be buried in the concrete or remain in the soil around the foundation; they contribute to subterranean termite infestations.

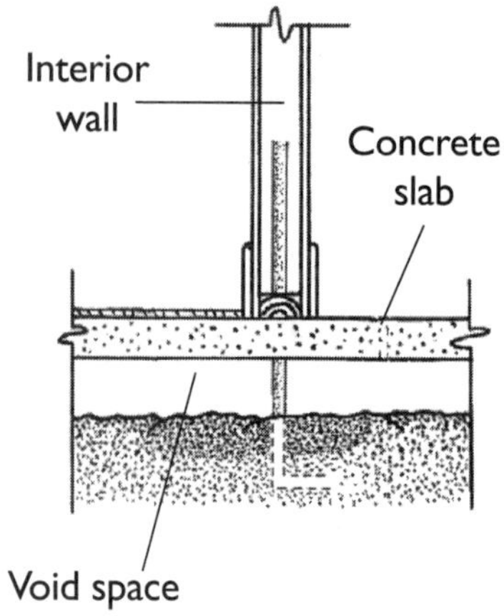

- *Concrete block.* Foundations are often made of concrete blocks 8, 10, or 12 inches wide, and the blocks will have two or three interior voids. Blocks start at the footing and are laid in an alternating configuration. Walls may be capped with solid block (cap block) or remain open, covered only by the width of the sill plate. The block wall for a crawlspace foundation is the same as that for a full basement.

  The outside surface of the foundation blocks is typically covered with a thin layer of concrete and also with a waterproof coating. These coverings would extend from the footer to at least the final grading of the perimeter soil.

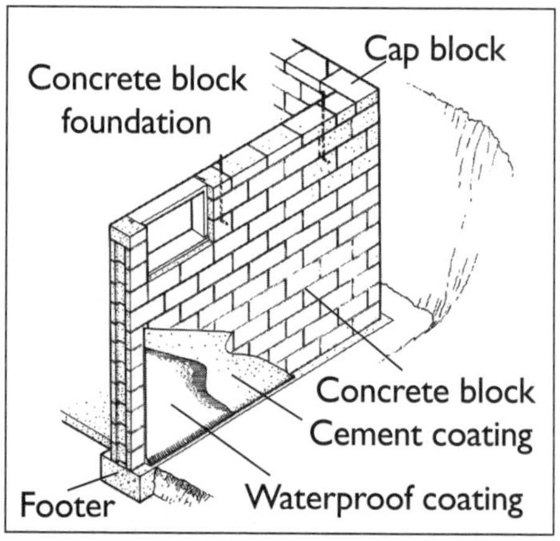

- *Slab on ground.* These foundations are common where soil conditions are stable. The combined slab and foundation consists of a shallow perimeter footing placed at the same time as the concrete slab. The bottom of the footing usually extends 12 inches below grade and rests on undisturbed soil. The slab is usually thickened under the interior load-bearing walls.

- *Crawlspace.* The crawlspace under houses may have high moisture and humidity that will promote wood-decay fungi and support various insects. Crawlspaces are usually ventilated and protected by a vapor barrier over the soil. Plastic sheets, such as 4- to 6-mil thick polyethylene, can be used as vapor barriers.

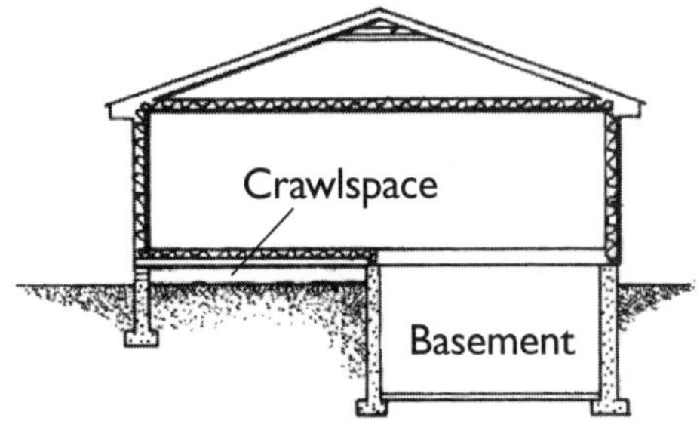

Most building codes require 1 square foot of ventilation for every 150 square feet of crawlspace, and 1 vent within 3 feet of each corner. The ventilation area may be reduced to 1 square foot per 1,500 square feet of crawlspace where the ground is covered.

Typically, 70% to 80% of the crawlspace soil surface is covered, leaving some soil exposure to provide minimum air humidity in the space and prevent shrinkage of sub-flooring and floor systems above the crawlspace.

Passive, non-electrical vents use a temperature-sensitive coil that is designed to open at about 70°F and close at about 40°F. Vents are usually covered with No. 8 mesh screen to deter insects. Powered vents have built-in electric fans to produce a directed airflow. Powered systems usually include a thermostat and a humidistat that monitor heat and humidity.

Rigid-Board Insulation. The outside of foundation walls may be insulated with rigid- or foam-board insulation. The boards are usually placed below grade and extend along the outside of the foundation wall to the footer. They may be hidden from view in the finished structure.

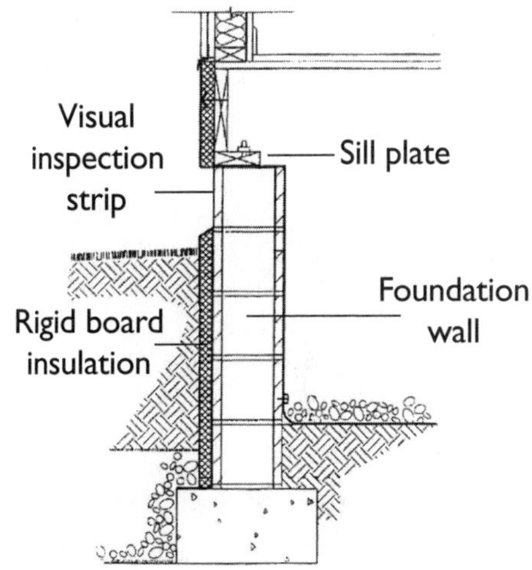

Rigid-board insulation that contacts the soil and extends below grade may create a pathway for subterranean termites from the soil to structural wood. Removal of the board insulation immediately above grade level will create a visual-inspection strip of about 6 inches around the perimeter.

ROOF COVERINGS

- *Asphalt shingles.* The most common roofing material is asphalt shingles. They are made from asphalt-impregnated felt, coated with a layer of colored stone or ceramic granules. Roof-underlay material usually consists of asphalt-based felt and plywood sheathing; this may not be used with wood shingles.

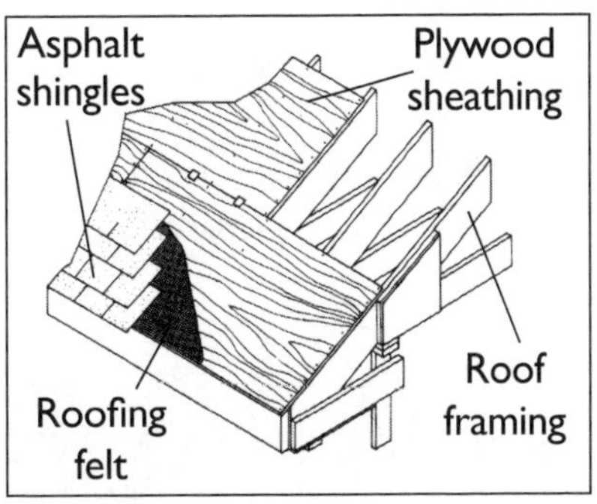

- *Wood shingles and shakes.* Wood shingles are usually western red cedar or white cedar, but redwood shingles are also used. Red cedar is a common shingle type, and it is highly impermeable to water. Wood shakes are longer and thicker than shingles and are made by hand. Shakes are from the same wood species as wood shingles, but they are laid on asphalt roof felt.

Asphalt shingles degrade over their service life, and water leaks develop and penetrate to the plywood sheathing and roof framing. Moisture-damaged wood can be infested by carpenter ants and termites.

Wood shingles are subject to infestation by carpenter ants and carpenter bees, in spite of the expected repellency of cedar.

The presence of decay fungi can be attractive to subterranean termites, and moisture-damaged wood can be infested by carpenter ants.

Exterior Wall Siding. Exterior siding may be damaged by moisture from wind-driven rain or moisture vapor from inside. This moisture may remain trapped between the siding and the building paper or other insulation material. It can lead to the development of mold and mildew, which will decay the material on both sides of the space.

- *Aluminum and vinyl siding.* This type of siding can be applied directly over wood-sided walls, insulated siding, or furring strips. It frequently contains thin foam inserts that may add insulating qualities to the aluminum.

  Vinyl siding is made of polypropylene or polyvinyl chloride resins. This material is resistant to ultraviolet light, but it is not waterproof and requires a weather barrier. It is applied as vertical or horizontal strips. This type of siding may appear wavy and reflect the irregularity of the underlying substrate, such as old wood siding.

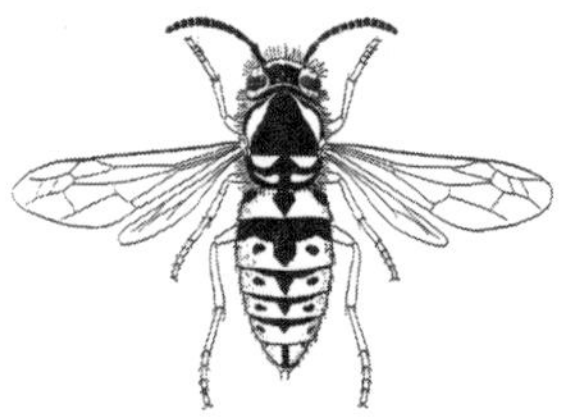

Yellowjacket

When vinyl siding is uneven, there may be sufficient spaces for yellowjackets and honey bees to enter. In the spring, queens may find these sites and establish nests. Nests often extend into wall voids and can be difficult to remove.

- *Stucco, cement.* Stucco and cement-plaster finishes are usually applied over a metal lath. Waterproof paper or a plastic-vapor barrier may be applied to the wood frame below the lath. Since stucco is applied over exterior structural walls, it is susceptible to cracking due to structural movement or stresses caused by changes in temperature and humidity.

- *Exterior insulation and finish system (EIFS).* This is a type of synthetic stucco, in which a plywood sheet against the frame forms the base for an insulating foam board that is attached to the plywood with an adhesive. A waterproof base coat is applied to the surface of the foam, followed by a finish coat of acrylic or co-polymer. Houses are sometimes covered with a synthetic stucco that does not have sufficient drainage, which, over time, can lead to moisture-damaged wood beneath the stucco. When extensive, such damage can be attractive to subterranean termites, and carpenter ants.

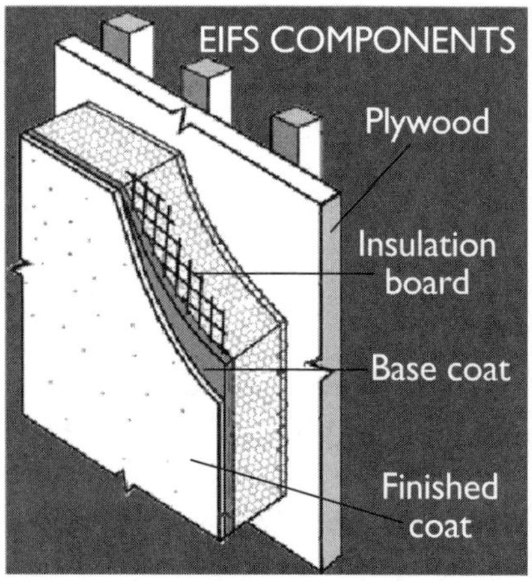

New EIFS products incorporate drainage ways and moisture-barrier membranes behind the insulation boards to allow for drainage.

- *Wood sheets, wood-overlap siding.* A number of wood species, including cedar, may be used to create overlap siding on houses. A vapor barrier is used between the siding and the frame. Without paint or treatment of any kind, cedar and other wood surfaces change color and texture in a few months, then stay almost unaltered for a long time. As weathering continues, all woods will become gray as the wood cells at the surface degrade.

- *Brick veneer.* Brick veneer may be a single width of brick, nominally 4 inches thick, or a thin layer of simulated brick material, which is about 0.5 inches thick. In full-sized brick veneer, the base brick is on the foundation wall and successive courses of brick are above this. The space between the brick and the frame is about 1 inch, and it may be partially filled with rigid insulation sheets.

Carpenter bees will nest in wood siding, including siding made of cedar. Solitary bees will establish nests in old wood-infesting-beetle holes in siding; bee-excavated frass on the outside of the hole can give the appearance that there is an ongoing beetle infestation.

When drilling through brick veneer to apply liquid insecticide, the narrow 1-inch spacing should be considered, and the liquid applied slowly to prevent splash-back and to permit time to inspect for both interior and exterior leaks.

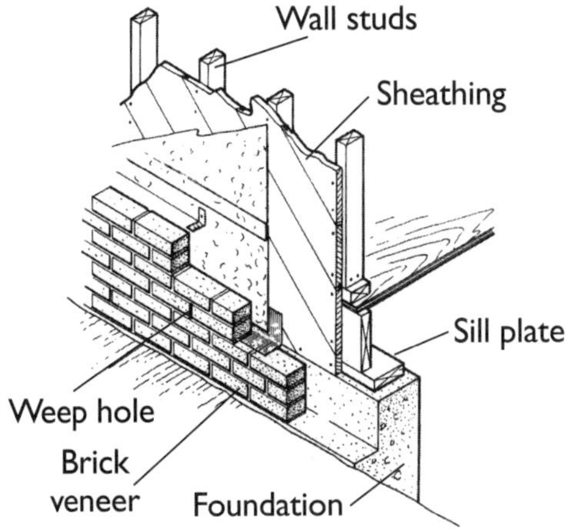

- *Brick cavity walls.* Cavity walls may be made of one or two load-bearing courses of brick with an internal void. Brick veneer may be placed on a concrete-block or poured-concrete foundation. When two courses of brick form the wall, an air space is maintained between the bricks. This space serves as a drainage gap between the two masonry units. Flashing at the second or third course of brick directs any moisture in the wall toward the outside.

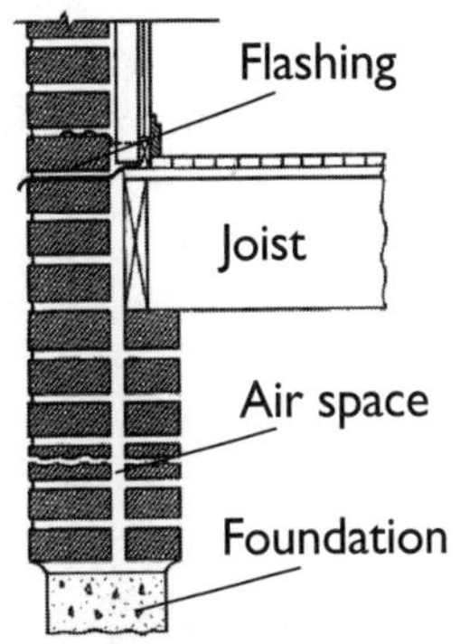

Weep holes may be open mortar space between bricks and contain a cotton wick; this allows for drainage of moisture in wall voids.

INTERIOR FEATURES

Heating, Ventilation, and Air Conditioning Systems (HVAC). There are two basic types of central heating systems for houses and small buildings: *furnaces*, which are used with ductwork for air distribution, and *boilers*, which distribute water or steam. Furnaces may be configured to deliver warm air in suspended ceilings, along the baseboards, or from units built into the floor or ducts buried in the concrete slab.

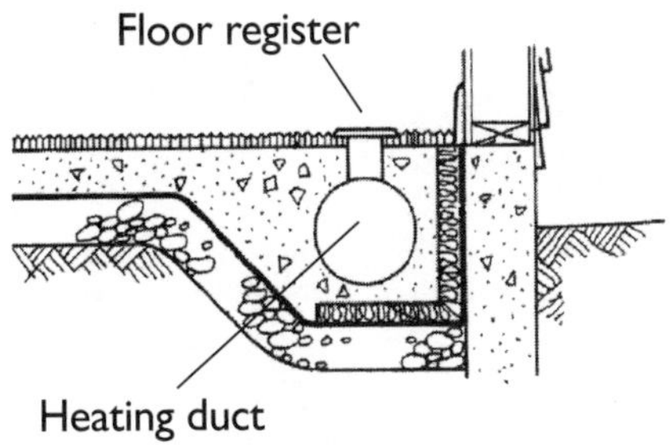

- *Radiant-heating systems.* Radiant-heating systems transfer heat from hot-water tubing or electric cables embedded in a concrete floor or ceiling. Household systems may use hot water through plastic pipes or metal tubing embedded in concrete slabs or framed floors. The tubing or cables may be laid in a grid or serpentine pattern within the flooring.

There are systems that run both hot and cool water through the same pipes. In some houses, heating pipes may be in the garage floor. This is important to know when drilling and treating this space for subterranean termites.

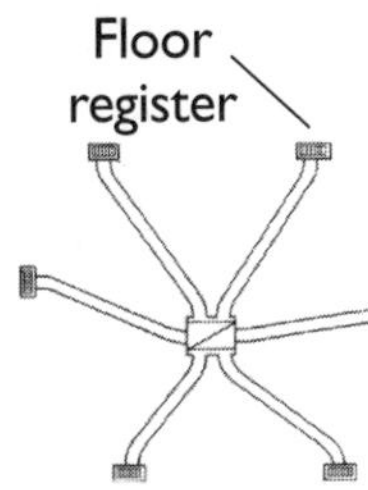

Concrete-embedded heating ducts, that are part of a perimeter-loop pattern for heating or cooling, are usually 2.5 inches from the surface. Heating pipes embedded in concrete slabs are also 2.5 inches from the surface. The spacing for grid pipes is typically 3 to 4 inches apart at exterior walls.

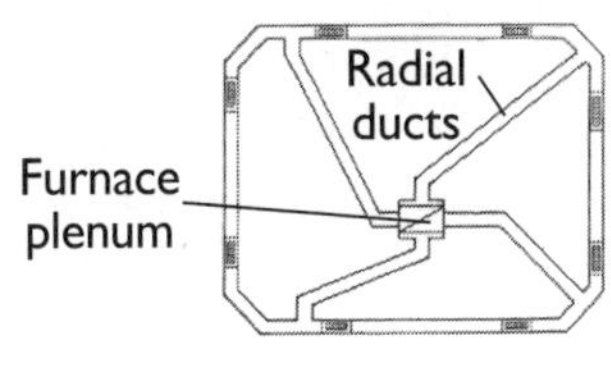

- *Ventilation.* The distribution pattern of the ventilation ductwork is variable and depends on the building structure. Concealed ductwork usually has horizontal

Termiticide labels require that the location and type of heat and air-conditioning ducts are identified and located before treatment to prevent contamination and damage to these structural elements.

elements that may be in the basement or crawlspace, or within the floor or roof.

Whole-house ventilators are typically installed in the attic floor over a hallway or stairway. They pull hot air from the living space and exhaust it into the attic where it may exit through the attic vents. A power attic or rooftop ventilator assists the natural flow of air through the attic space.

Sources of Water in Buildings. Humidity greatly influences household substrates, such as wood, leather, and organic fabrics. Excess moisture in these materials may cause swelling and distortion, and, in some cases, enhance the growth of molds and fungi.

- *Relative humidity (RH)* is an expression of how much moisture air is holding compared to how much it could hold at that temperature. For example, an RH of 80% at 80°F means that air is 80% saturated with water. This is not equivalent to an RH of 80% at 50°F, because, as the temperature of the air falls, its capacity to hold water decreases.

  The presence of free water in buildings can influence the introduction and survival of mold, fungi, and wood-infesting insects. This water can be absorbed into structural wood and support mold on the surface, decay fungi below the surface, and ants excavating nest sites within. The sources of this water may be interior leaks and standing water or condensation on pipes and structural wood in attics and basements.

- *Condensation.* The first sign of condensation problems in a building is the appearance of mold growth on interior surfaces, often near the top of external walls.

  - **Windows**. Condensation and mold growth can be widespread or be in scattered patches. Favorable areas for condensation include corners of external walls, and cupboards and closets built into exterior walls. Another site is the junction of the glass and window frame of single-glazed windows, where water often accumulates. In double- or triple-glazed windows, the metal frame forms a thermal bridge to the cold exterior, and condensation does not occur.

At window locations, condensation provides conditions favorable for carpenter ants and subterranean termites. Mold growth here can provide conditions and food for psocids, springtails, and other pests.

  - **Basements and crawlspaces.** Throughout the year, the temperature of the ground surrounding basements is lower than that indoors. The relative humidity during summer may be high enough for condensation to appear on exposed wooden surfaces. The temperature difference between the basement and the occupied space above it may contribute to wet surfaces and mold growth.

    In crawlspaces under floors, evaporation from the soil can produce high humidity in the air. Floor joists and floorboards take up water from sub-floor air, and this can increase wood moisture content to 20% or more. High wood-moisture content often occurs in crawlspaces where the above floor is covered with vinyl or ceramic tile. Another source of water in crawlspaces is drainage from air-conditioning systems.

    Wood moisture above 20% is suitable for the growth of wood-decaying fungi.

Wood that has sustained high moisture content can be attractive to subterranean termites and other wood-infesting insects.

- *Water-vapor barrier.* Polyethylene sheeting is used in crawlspaces as a vapor barrier to limit the movement of water in the soil into the crawlspace and the exposed wood. Usually 70% to 80% of the soil surface is covered, allowing some moisture in the space to keep wood from shrinking. Polyethylene is not impermeable to water vapor, however, so moisture is known to move through the material.

- *Roof spaces.* Modern houses are designed and often refitted to be draft proof or sealed, and the outside covered with vinyl siding. This leaves the roof space as the only area for water vapor in the house to escape. Water vapor can pass up through ceilings and insulation into the roof space where it may condense on roof timbers.

- *New buildings.* Water evaporates from green timber, concrete, gypsum plaster, and other materials used in new buildings. The rough frame of a house may be damp when constructed, and rain may add more water before the sidewalls and roof are completed. Because of rapid completion of the building, this moisture may remain for months after occupancy. An average of 3 gallons per day is released into the indoor air during the first year; this is reduced by half the second year and is negligible thereafter.

  The "new-construction pests" that feed on fungi growing on the surface of wet wood, plaster, and other materials are psocids, foreign grain beetles, silken fungus beetles, and minute brown scavenger beetles.

- *Flooding.* The outcome of kitchen and bathroom overflows and plumbing breaks is extensive, and sometimes sustained, water exposure of framing timber and wall surfaces. The result is some degree of bacteria, mold, and fungal growth; drying wet surfaces within 24 to 48 hours can help to prevent this growth. In the case of wallpaper, the moisture content should be decreased to about 7%, and the surface of structural wood should be less than 20%. The goal of book and paper-product restoration is 5% to 7% moisture content.

- *Damaged wallboard.* This material provides a source of cellulose on which fungi and other microorganisms can feed, and subterranean termites can use as a food source. Plasterboard is made by sandwiching gypsum between cardboard layers. Relatively little water is required to raise the moisture content of the wrapper to a level that supports mold growth.

Household Pets. Household pets now extend beyond the typical dog or cat to include a wide variety of birds; rodents; reptiles, such as snakes and lizards; rabbits; and arthropods. Depending on the type of pet, owner, and household conditions, these animals may be confined to one location or have access to nearly all rooms. Their movement indoors can influence the materials and methods used for control.

Before indoor spraying of insecticide, birds should be removed or the cage completely covered, and reptile and arthropod cages completely covered.

The influence of house pets is the addition of hair, feathers, and skin scales to the habitat. This material supports infestations of dermestid beetles, dust mites, and other pests.

Floor, Sink, and Shower Drains. Floor drains provide food and breeding sites for pests such as flies and cockroaches. The most common flies that breed in continuously wet floor drains are moth, or drain, flies and dark-eyed fruit flies. Fruit-fly larvae are usually in the decaying material around the top grate of the drain opening; moth fly larvae are usually in the organic material below the opening, sometimes in the sediment trap.

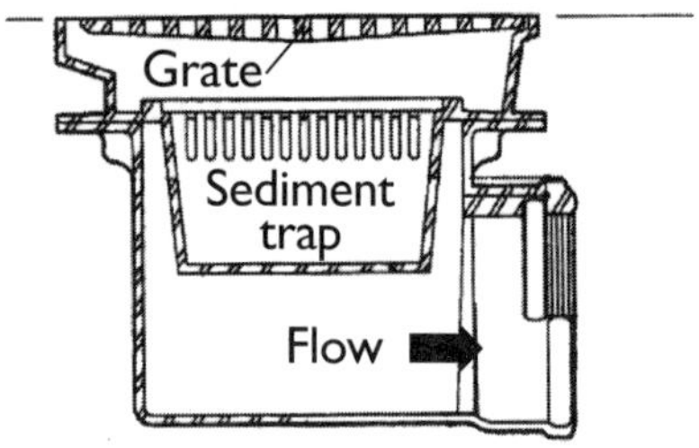

Flooring drain

Smooth sides of bathtubs and sinks often trap oriental cockroaches, silverfish, and house centipedes, and give the impression that they have come up through the drainpipe. Bathroom shower drains present a special problem because of the potential of organic matter, such as soap, shampoo residue, and hair, accumulating around the grate and in the trap. These drains may go for long periods without being cleaned and become breeding sites for moth flies.

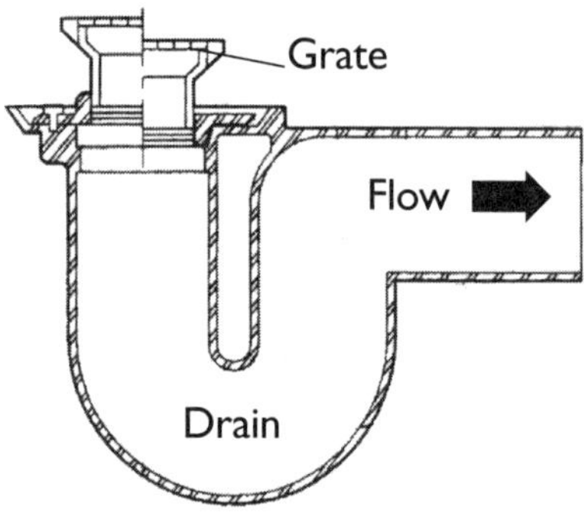

Shower drain

Fabric. Clothing made from natural or synthetic fibers and leather stored in quantities indoors is often unprotected from pests. The larval stages of clothes moth and dermestid beetles are capable of digesting the natural fibers and fur. Although this substrate may be their preferred food, they will attack cotton, linen, silk, and synthetics when there are stains or residues of nutritional value.

- *Wool.* Wool has some resistance to bacteria and mildew, but these organisms will attack and destroy fibers stained with organic matter and soil. Carpet beetles are the most damaging pests of wool. Clothes moth caterpillars are capable of digesting wool and will feed on fabrics of vegetable origin or synthetics if the fabric is mixed with wool.

- *Silk.* Silk is resistant to mildew, bacteria, and fungi. It has some resistance to clothes moth larvae, but not to the feeding of carpet beetle larvae.

- *Cotton.* Clothes moths and carpet beetles will not attack cotton, unless it is soiled with organic substances. Silverfish will eat cotton that is starched. Termites will attack and feed on cotton products.

Moth fly larvae are immersed in organic matter while feeding, with only the tip of the narrow breathing tube on the posterior end in contact with the surface; breathing is from back to front.

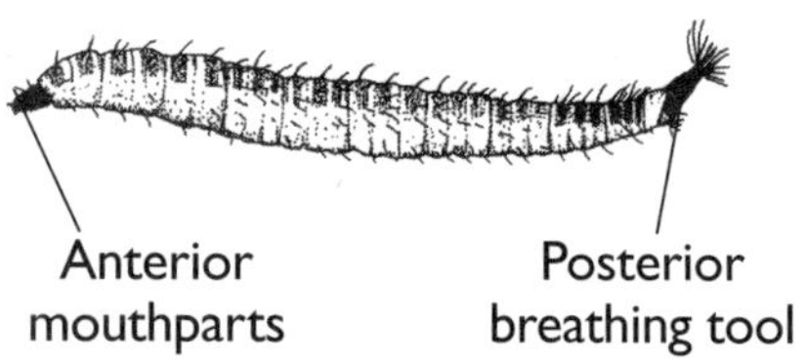

Moth fly larva

Floor Coverings. Floor coverings for interior surfaces include hardwoods and softwoods, sheet flooring, clay and ceramic tile, and carpeting. In old or historical houses, the use of softwood in floors may be extensive. Solid-wood flooring will shrink or contract during periods of low humidity when heating systems are running, leaving cracks between floorboards. Flooring will expand during periods of high humidity.

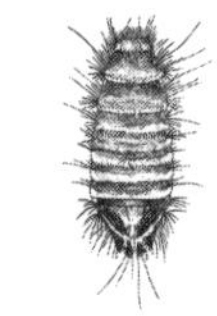

Varied carpet beetle

- *Carpeting and rugs.* Carpeting contains most of the house dust mites and mite allergens in houses and other buildings. Natural fibers can absorb considerable amounts of moisture from the air, and this provides a stable environment for dust mites. The cut and shape of the carpet pile contributes to the attack by insects and penetration of insecticides. Carpeting may be categorized as straight or twisted pile; cut-, loop-, or cut-and-loop pile; and pile or woven.
    - **Cut pile** carpet has long and twisted strands that usually do not stand upright. Dirt penetrates this carpet easily, and the loose strands make it difficult to treat with insecticides.
    - **Loop pile** carpet has no exposed tips; it has some resistance to dirt and liquids.
    - **Ribbed carpet** is a cut-and-loop style, with a combination of strands that are looped and some that are cut.
    - **Plush and velvet** carpets have short, uniform, and dense pile; they resist dirt and penetration of liquids.
    - **Wool and silk** are the animal-based fibers for carpets and rugs, with some being a blend of silk and wool fibers. These carpets are susceptible to insect infestations, with carpet beetles and clothes moths readily feeding on the animal-based fibers.

- *Hardwood strip and wood block.* Hardwood-strip flooring has tongue-and-groove edges and ends, and the bottom surface is hollowed. Strips are usually laid at right angles to the floor joists and supported with a plywood sub-floor. Strip flooring over a concrete slab is attached to a wood support frames, called sleepers. Polyethylene film may be placed between the flooring and the sleepers. Beneath the concrete slab may be another vapor barrier and a void due to settling of soil.

Block flooring consists of rectangles or squares with tongue-and-groove edges. It may be in the form of parquet flooring, made up of small sections of hardwood, or laminated flooring, which consists of bonded layers of hardwood. This type of flooring is usually laid over a thin coat of glue applied to a layer of asphalt felt paper, which is held in place over a concrete slab or sub-flooring by an adhesive.

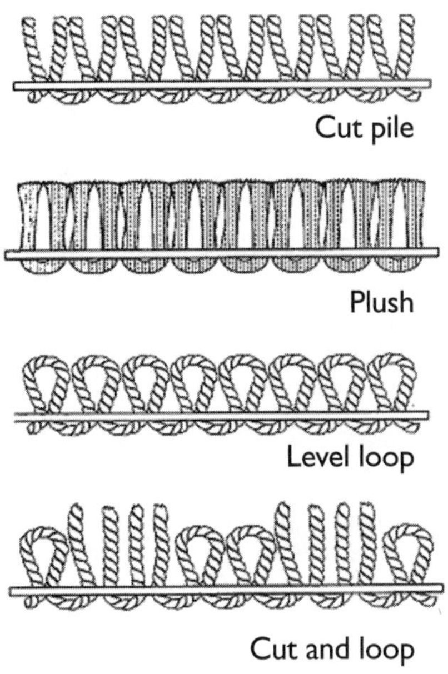

- *Tile.* Block tiles of linoleum, asphalt, vinyl, cork, and rubber are placed on plywood, particleboard, or hardboard. The thickness of the sub-flooring or underlayment boards depends on the room; kitchen floors may be thicker than floors in other rooms. When laid on a concrete slab, a vapor barrier may be placed between the concrete and flooring.

  Ceramic and clay tile may be installed on a concrete slab with a layer of cement as a base coat. This style of flooring may be installed above grade and with a plywood sub-floor.

- *Sheet flooring.* The vinyl sheet flooring used in modern houses is sometimes referred to as linoleum, but it differs from that material. Linoleum contains cork, wood flour, and ole-resins, while resilient sheet material contains vinyl resins, such as polyvinyl chloride, with a fiber back.

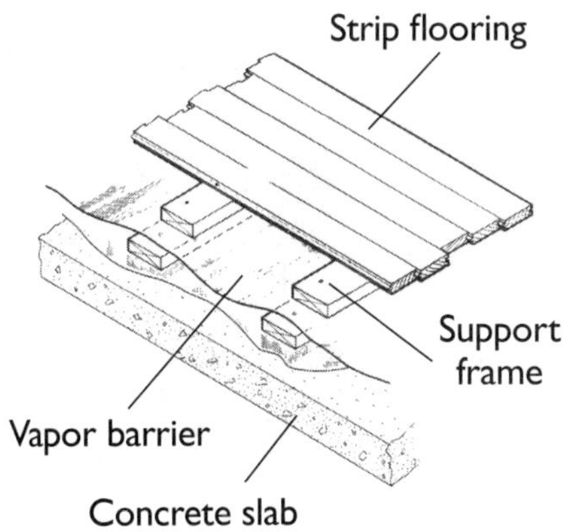

CHAPTER 11

TERMITE CONTROL

Modern termite control considers each site unique, with each being treatable with a combination of strategies, such as liquid application and baits. Effective termite control also includes non-chemical materials and methods. Most of the current termiticides are formulated as liquids for below-ground application. The objective of soil treatment is to establish a continuous zone of treated soil that will intersect with termites foraging around a structure. Baits provide a toxic material that is returned to the nest and eventually eliminates the colony. Non-chemical strategies include barriers, pliable membranes, extreme heat, and cold.

All termites are not alike. There are differences in biology and behavior among closely related species, even those occurring together in the same geographic area. The species that attack structural wood are classified as subterranean, dampwood, and drywood termites. Subterranean termites nest in the soil and require a source of moisture for survival and feeding. Dampwood termites are less common, and they infest wood that is wet and decayed by fungi. Drywood termites infest wood that is not in contact with soil, and they depend on the moisture in the wood they infest.

SUBTERRANEAN TERMITE BIOLOGY AND HABITS

The several species of subterranean termites that occur in urban and suburban areas are major structural pests. There are few regions that are free of subterranean termites. The natural reservoirs for these termites may be vacant lots, small wooded areas, and untreated buildings. Modern termiticides act to prevent and control existing infestations and eliminate colonies.

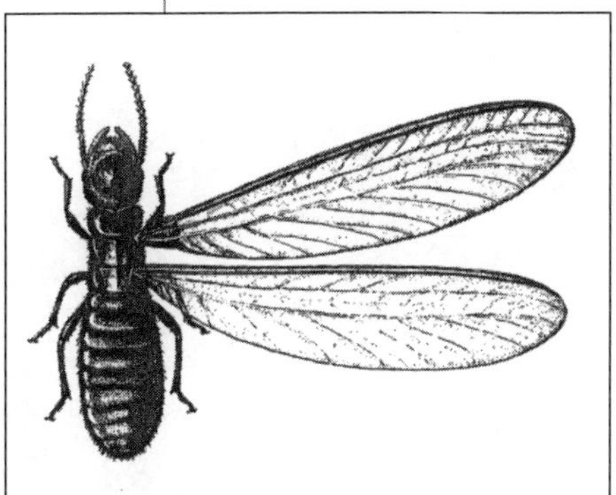

Subterranean termite

- *Distribution.* There are several termite species that are native to North America. Species in the eastern U.S. include *Reticulitermes flavipes*, *R. virginicus*, and *R. hageni*, with the occurrence of these three species overlapping in the Mid-Atlantic States. *R. hesperus* is common in the western U.S. The Formosan termite is an introduced species that has a limited distribution in southern states.

- *Infestations.* Buildings can be infested with several distinct colonies of subterranean termites. There may be as many as seven colonies in the soil surrounding some infested structures. The cause for termite control failures or continued infestation may be due to the presence of multiple colonies. When several colonies are present in a vicinity, the elimination of one may cause another to take its place in a foraging area or infested structure.

Colony Structure and Activity. The concept of an organized colony may not be completely accurate for subterranean termite species of the U.S. Most have a colony structure that consists of a number of feeding sites and probably less than 100,000 individuals. The traditional enlarged queen is not common in these colonies, rather eggs are often produced by supplementary queens.

- *Shelter tubes.* Subterranean termites move between the ground and a source of food by building earthen shelter tubes. Workers build tubes from particles of soil and feces, bound together with secretions. Workers can build tubes at the rate of 3 to 4 inches per 24 hours. There are different types of shelter tubes:

- **Utility tubes** are the main connection between above and below ground portions of the colony;
- **Exploratory tubes** are often free-standing and extend up from the ground, but do not reach the wood above;
- **Drop tubes** extend downward from infested wood.

Workers rebuild tubes three times faster than they build new ones; broken tubing is built at the rate of 1.2 inches per 65 minutes.

- *Foraging.* This is the primary activity of the colony; it occurs throughout a network of tunnels and is not random. When termites tunnel through a portion of their territory and discover a food source, the foraging pattern changes. Their tunneling radiates out from the food source until another food source is located, and the pattern starts again. Once food is located, workers shift their tunneling from a searching mode to a concentrated-direction mode.

- *Seasonal foraging.* There is a seasonal cycle of activity in colonies, with most of the foraging occurring during summer and early fall, and less during winter. Natural fluctuations in colony feeding influence a baiting program. Baits may affect termite activity faster during summer months than winter. When bait treatments are conducted during the summer, reduced termite activity at the site is typically seen during the first three months of bait acceptance.

- *Food exchange and grooming.* Colony workers groom each other with their mouthparts. This cleaning removes microorganisms from the cuticle and spreads pheromones that regulate colony development. The mutual exchange of food between nestmates is called "trophallaxis." The transfer of partially processed food occurs from mouth to mouth or anus to mouth. Trophallaxis can also circulate insecticides throughout a colony.

Termites that ingest a pesticide from a bait station will spread it to nestmates by trophallaxis.

- *Swarming.* Subterranean termite colonies usually produce swarmers after about five years. The number produced is in direct correlation to the size of the colony. Most colonies of subterranean (*Reticulitermes*) species produce several large swarms in the spring, but swarms may occur in heated buildings in mid-winter and continue into spring.
  - ***R. hageni*** swarms in late summer and fall in the southeastern U.S.
  - ***R. tibialis*** swarms during winter or spring, and again in the fall.
  - **Formosan termites** swarm from dusk until midnight from April through July.

Swarming occurs when the temperature has been within a certain range for a period of time, and rainfall occurs near the end of this period. The initial swarm usually occurs within three days of the first rainfall. In the absence of rainfall, the colony can delay swarming for weeks.

Termite control that occurs immediately after adults are produced in a colony may not prevent swarming. This can give the false impression that the control was ineffective.

New Colony Formation. New termite colonies may be created through:

- *Swarming.* The release of large numbers of winged adults from a colony is one way new colonies are formed. Males and females from the original colony or from other colonies form mating pairs. This queen and king will establish a nest site and begin production of eggs and worker termites. However, swarming as a method of colony reproduction is not very effective; few of the pairings are successful, and few result in new colonies.

- *Budding.* The most common method of colony reproduction is the splitting of a large colony into one or more smaller units, this is called "budding." Over time colonies can occupy a large and often irregularly shaped territory. Portions of the colony become isolated and respond by producing secondary queens, which are capable of laying eggs. These small satellite units may be separated from the original colony and soon begin functioning as an independent colony. Separation may be the result of physical disruption of the nest site, or application of termiticide that separates nestmates in the soil from those in the structure.

- *High rise.* In some populated areas, subterranean termite colonies may become established on the flat roofs or upper floors of high-rise buildings. In these locations there is little or no soil associated with the colony. But, because there is a sufficient supply of water and moisture-damaged wood in these locations, the colonies are often successful and exist for many years.

Wood Consumption. Termites show distinct feeding preferences: they prefer softwood to hardwood, fungus-decayed wood to sound wood, moist wood to dry wood. All of these preferences influence the amount of wood eaten by a colony. Estimating wood consumption by a colony is problematic because it is influenced by conditions, colony size, and probably other factors.

Under ideal laboratory conditions with no alternative food available, 60,000 subterranean termites can eat 0.2 ounces of wood per day, which is about 1 foot of a 2x4 board eaten in 157 days.

FORMOSAN SUBTERRANEAN TERMITE BIOLOGY AND HABITS

The Formosan termite (*Coptotermes formosanus*) is more vigorous and aggressive than other subterranean species in North America. There is extensive tube and tunnel building to locate a food source; colonies are larger than the native termite species; and it can cause structural damage within a few years. This species can also be a pest of living trees.

Distribution. This species occurs along the Atlantic Coast from North Carolina to the Gulf Coast region; and in Texas, Louisiana, Mississippi, Alabama, Florida, Georgia, South Carolina, Tennessee, and in San Diego, California.

Colony Structure and Activity. Formosan termites nest in soil, but also construct nests above ground with no contact with the soil. These infestations require a source of moisture, such as plumbing leaks, roof leaks, and condensation on pipes. The actual nest is composed of "carton," partially digested wood cemented together with salivary secretions and feces. The walls of the carton nest are porous and hold moisture, and the interior has a network of galleries. Formosan termite colonies are significantly larger than other subterranean termite species.

The average mature Formosan colony contains 350,000 to 2 million individuals, while the colony size of Eastern subterranean termites contains 60,000 to 1 million workers.

Swarming. Formosan termites swarm between dusk and midnight, from April through July. Large numbers can gather around street lights and other outdoor lighting. Swarms may also occur indoors.

Drywood termite frass

DRYWOOD TERMITE BIOLOGY AND HABITS

The most common drywood termite species in North America are the Western drywood termite, *Incisitermes minor,* and the Southeastern drywood termite, *I. snyderi*. The West Indian drywood termite, *Cryptotermes brevis*, occurs in Hawaii, Florida, and some coastal cities in the southeastern U.S.

Colony Structure and Activity. Infestation sites include lumber in sub-flooring, walls, attics, and roofs. A colony begins with a single king and queen and develops over several years. These termites do not live in the soil and do not require contact with soil. Colonies of the Western drywood termite develop slowly. A colony may contain several thousand individuals and survive for about 10 years. These termites consume a relatively small amount of wood. A colony containing 1,000 termites would consume about 0.5 pound of wood per year.

Fecal Pellets. Hardened fecal pellets are extruded from holes in the infested wood; they may accumulate in piles directly below the holes. They have longitudinal ridges, one rounded end, and one tapered end. Pellets are pale white to dark brown, but the color would depend on the wood infested and, to some extent, the age of the pellets. Vibrations cause pellets to fall from wood after control; indoors, the pellets may be scattered from the original "kick-out" hole.

Swarming. These termites swarm over a period of several weeks. The Western drywood termite swarms during sunny days in fall, the Southeastern and West Indian drywood species swarm at night from April to June.

New Colony Formation. New colony formation depends on the pairing of winged adults and the location of a suitable nest site. Following the mating flight, the male and female seek a point of entry to wood, such as under eaves and roof overhangs, and through joints between rafters and ridge plates. Colonies can infest wood on the upper floors of buildings. Infestations may remain unnoticed because there are no earthen tubes as evidence, and colonies are usually small. Infestations may spread from galleries in the wall frame to furniture, and when furniture is moved to a new location, the infestation may spread from furniture to structural wood.

DAMPWOOD TERMITE BIOLOGY AND HABITS

Dampwood termites require humid to wet habitats; their nests are located in damp and decaying wood. The Pacific dampwood termite, *Zootermopsis angusticollis*, causes damage to log houses and utility poles along the Pacific Coast from British Columbia to Baja California. *Neotermes jouteli* is found along the coastal region of southeastern Florida, and *N. castaneus*. *Paraneotermes simplicicornis* occurs in southern Florida.

Colony Structure and Activity. Dampwood termites establish nests directly in wood rather than in soil. Development is slow, and it may take several years for a colony to have 4,000 workers. The maximum size colony for some species is only about 1,500 individuals. Swarming occurs in late summer and fall after rainfall, with the termites attracted to outdoor lights.

Fecal Pellets. The fecal pellets of the Pacific dampwood termite are slightly hexagonal in cross section. They lack the longitudinal ridges that are found on drywood termite pellets.

Swarming and New Colony Formation. These termites swarm in late afternoon and evening, often after rainfall. Most colonies swarm between August and October; some species swarm in the spring. New colony formation for drywood termites depends on the successful pairing of adults.

INSPECTION FOR TERMITES

Visual inspection of structures by trained technicians is the most common method of detection and location of termite infestations. However, there are structural features that can limit this method. Generally, only 25% to 35% of the inside of a modern house is accessible for inspection. Based on re-treatments and re-inspections, 40% of subterranean termite infestations originate at utility entrances through slabs, and settlement and expansion cracks in slabs.

- *Swarmers.* The most common indication of a subterranean termite infestation is the sudden indoor emergence of a large number of winged termites. The swarm can take place outdoors and go unnoticed or be confused with winged ants. These termites are attracted to electric lights or windows, or they may simply crawl on the floor and walls. Within a few hours they shed their wings, and most die.

- *Termite wings.* Adult male and female swarmers shed their four wings soon after emerging from the colony. When swarming occurs indoors large numbers of these slender, silvery wings will collect on the floor, in light fixtures, or at windows. The sudden presence of a large number of shed wings indicates a recent swarming and the likelihood of an infestation. Termite wings may remain in a structure for a long time; they do not decay. The presence of a mass of adult insects and shed wings can also cause allergic reactions in some people.

- *Shelter tubes.* Earthen tubes extending from the soil to above-ground wood are usually attached to a foundation. They may be visible on poured concrete walls or concealed

Subterranean termites

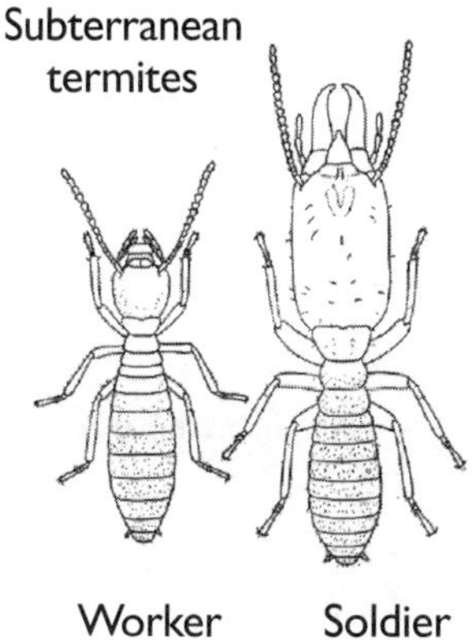

Exploratory tubes

The surface of wood infested with drywood termites can show little evidence of galleries. There may be small piles of fecal pellets along infested wood.

in the voids of block walls. Shelter tubes may be built across the gap between a concrete slab foundation and the wood framing above. Exterior insulation boards that extend below grade can conceal termite tubes built against the foundation or within the insulation.

Shelter tubes on exposed surfaces should be removed during or after treatment. They do not crumble in the absence of termites and may give the incorrect impression of an active infestation at a later inspection.

- *Wood surface.* The surface of infested wood may show signs of blistering or weakness if subterranean termites are feeding in the outer layer. Often tunnels may be confined to the interior of a timber, with the surface showing no signs of weakness. Probing the surface of wood with a screwdriver, pick, or drill can indicate termite galleries below, because the surface of infested wood is thin and can be penetrated with minimum pressure.

- *Termite galleries.* The feeding galleries of subterranean termites are lined with a thin layer of soil. The galleries are open and may extend for a long distance in a piece of wood. Termite galleries do not contain powdery frass, but the galleries of drywood termites may contain small amounts of granular fecal pellets.

- *Termite workers.* When there is an active infestation, shelter tubes and galleries in wood contain termite workers and soldiers. Workers may be absent if there has been a disturbance, or activity has moved to another location. This also applies to shelter tubes that are removed and not rebuilt; it may simply indicate a shift in feeding sites or the replacement of one colony with another.

Location of Infestations. Locations around the outside and inside of a structure that are most conducive to subterranean termites include those that provide moisture and wood-to-soil contact. However, there are many features of buildings that provide access for termites; openings as small as 1/32 inch can be used by foraging workers.

- *Attached slabs and stairs.* Concrete slabs and stairs attached to a structure can be the starting point for termite infestations. These slabs may develop cracks that extend to the soil beneath, or there may be a sub-slab void beneath the concrete; often there is also an expansion joint at the junction of the main structure. The fill dirt beneath these slabs may contain construction paper and wood scraps, which are a food source for termites.

- *Crawlspaces.* High humidity and the exposure of the sill plate, joists, and subflooring in crawlspaces provide conditions for termites. Problems with crawlspaces include proper drainage on the exterior and the absence of vents. Crawlspaces are often narrow and not easily inspected, but because there is high potential for termite infestations to begin there, inspection is required.

- *Foundation wall and sill plate.* The sill plate may be at the top of a solid concrete hollow-block foundation. The top course of block may be open to the footing, or it may be capped with a half-size solid block (cap block). Block foundations

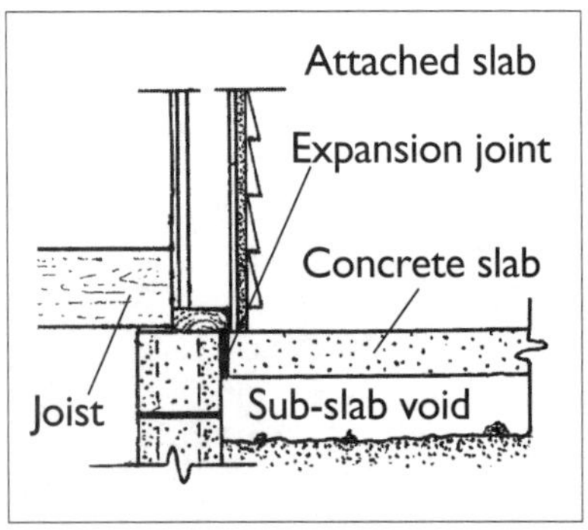

provide opportunity for termites to move from the soil to the sill plate without detection. Solid concrete walls prevent this, but shelter tubes may be built on the wall surface. Regardless of the foundation type, the sill plate is vulnerable to termites and requires careful inspection.

- *Utility pipes, wires, and cables.* Foundation walls, concrete slabs, and basement slabs are often penetrated by utility pipes, electric service wires, and various cables. The seal around the opening to the outside of these features may be incomplete or fail altogether, providing access to termites.

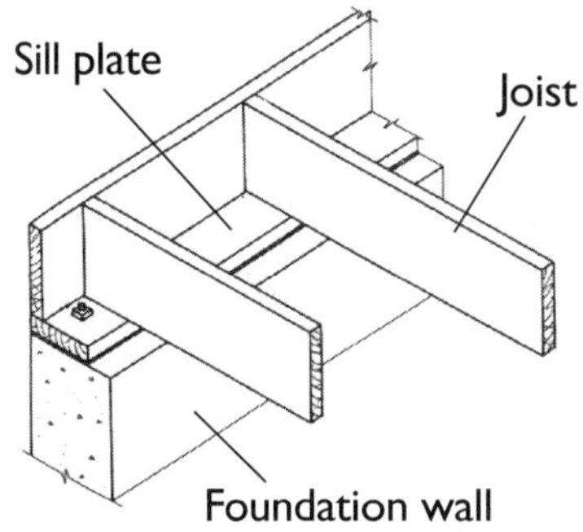

TERMITICIDE APPLICATION

Soil treatment remains the most common termite control strategy. Most termiticide labels require establishing a continuous and uniform chemical barrier in the soil surrounding the structure.

Termites tunnel in soil by following roots and the natural edge between soil layers. As they tunnel, they can penetrate 1/8-inch gaps in the treated soil surrounding a building.

Application Volume

- *Soil surface application (horizontal treatment).* Soil surface treatment is 1 to 1.5 gallons per 10 square feet of soil surface. Label directions require termiticide to be evenly and uniformly applied over the entire surface area.

  - **Square feet** is determined by multiplying the length by the width of the area to be treated. Do not estimate; small differences in length or width can result in large differences in total square footage.
  - **Linear feet** is determined by adding the perimeter lengths of all sides of the structure. A building 56 feet long and 24 feet wide has 160 linear feet.

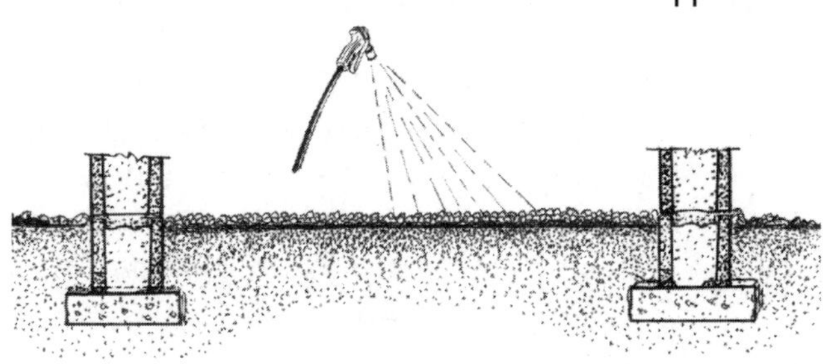

- *Below-ground application (vertical treatment).* For soil rodding, the recommended volume is 4 gallons per 10 linear feet per foot of foundation depth. Rod spacing is limited to 1-foot intervals, and trenches must be a minimum of 6 inches deep. The maximum foundation depth required for treatment is 4 feet.

  The applicator must determine the liquid delivery time for each rod insertion. Time is based on the 2-gallon-per-minute flow rate of the nozzle on the soil rod.

Delivery times for foundation depths are:

- *1 foot: 12 seconds*
- *2 feet: 24 seconds*
- *3 feet: 36 seconds*
- *4 feet: 48 seconds*

Clay soils may not accept the standard volume of liquid. Some product labels permit mixing twice the concentration of active ingredient and applying half the volume.

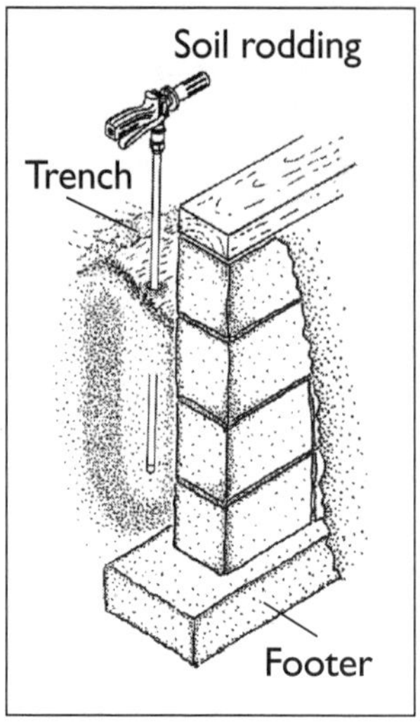

Sub-Slab Injection (Vertical Drilling). For treating under a concrete slab, holes are drilled through the slab for insertion of a treatment rod. Application is along the perimeter, adjacent to the foundation wall. Drill holes are spaced 1 foot apart and 1 foot from the wall, and liquid is applied into the void beneath the slab. A 180° or 360° directional tip will direct the spray toward the adjacent wall and footer.

Sub-slab injection

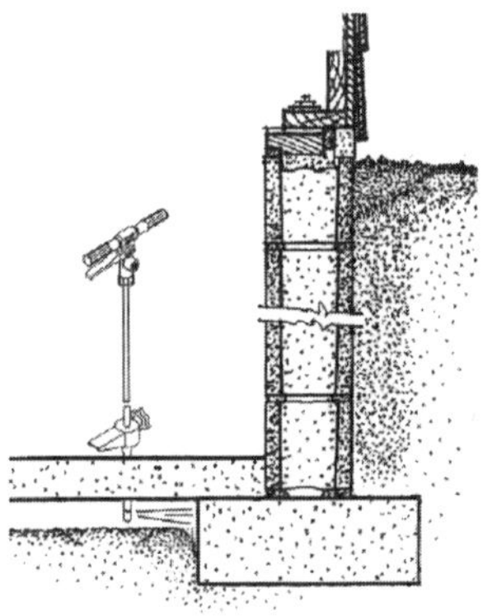

Hollow-Block Walls. Treatment of hollow-block foundation walls creates a treated zone in the voids and adjacent to the footing. Blocks may be drilled if there is no accessible opening at the top (e.g., if it is sealed with a cap block). Application is 2 gallons of finished dilution per 10 linear feet; holes should be drilled as close to the footing as practical.

APPLICATION EQUIPMENT

The valves, rods, and nozzles determine the amount and spread of termiticide in the soil. A 180° or 360° directional tip on the rod assures the continuous spread and overlap of liquid in the soil.

Valve. Professional-use valves have a balanced on/off trigger which allows the operator to control liquid delivery while inserting or withdrawing the rod with both hands. Valves with control of liquid flow built into the handle are important for long treatment times and efficient application.

Spacing the drill holes too far from the wall or too far apart will leave gaps in the treatment zone.

Soil Rods. Soil treatments and sub-slab injections are done with pipes (rods). The length of the rods is based on the 4-foot (48-inch) maximum treatment depth required on product labels. Rods are typically 5/8 inch in diameter and 40 inches long. When used in a 6-inch trench, with a 1-inch nozzle, and coupled with a 3- to 6-inch delivery below the nozzle, the 48-inch treatment depth is achieved.

Slab injection rods are typically 7/16 inch in diameter and 40 inches long. A flexible cone and pedal at the bottom of the rod seal the drill hole during application and prevent backsplash of liquid to the surface of the slab.

Sub-Slab Injector. Application beneath interior slabs covered with tile, wood flooring, or other material can be done with the sub-slab injector. The low-profile design includes a flexible gasket that expands and completely seals the 9/16-inch drill hole around the ½-inch pipe. Various lengths of application pipes are available to extend through sub-flooring to reach the void beneath concrete slabs.

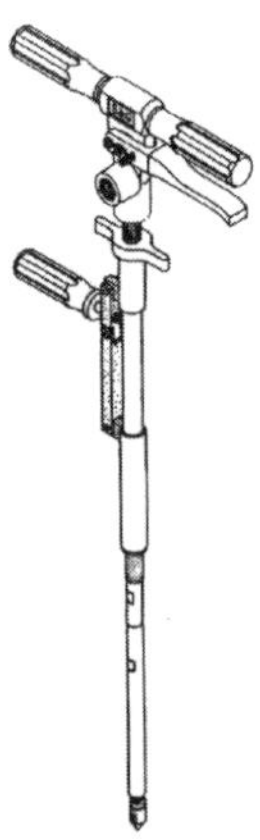

Sub-slab injector

Nozzle. Termiticide application requires directional nozzles to create an effective treatment zone in the soil. A 2-gallon-per-minute delivery rate and spray pattern of a 180° or 360° tip will direct liquid toward the foundation during soil treatment or toward the foundation wall when treating beneath a slab.

- *Directional nozzle.* The spray patterns of the 180° and 360° nozzle are essentially the same. The 360° tip is used for perimeter soil treatment. Under the pressure of 25 psi, the lateral spread is about 12 inches on both sides of the nozzle. This spread matches the 1-foot spacing required on termiticide labels to create continuous and uniform coverage in the soil. The depth of the treatment is about 16 inches, depending on the soil type. The horizontal and vertical spread is greater in sandy soils than clay soils.

- *Open or straight nozzle.* The open or "straight" tip has no spray pattern because the liquid flows directly from the end of the rod; there is minimal lateral spread during treatment. The dispersal of a straight tip is only about 6 inches on both sides of the injection site, which will leave an untreated gap when rodding at the required 1-foot spacing. Although the depth of treatment may be greater than the directional nozzle, the actual spread at the lower depths is limited.

Directional tip pattern

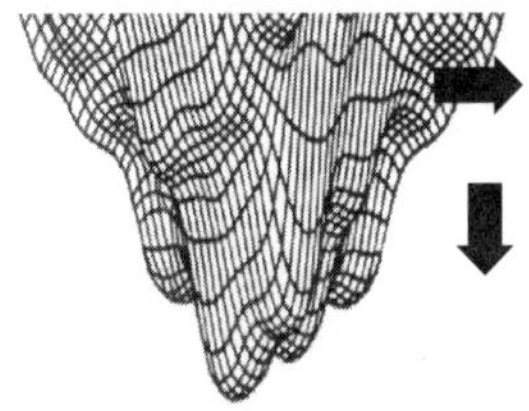

12" wide
16" deep

Straight tip pattern

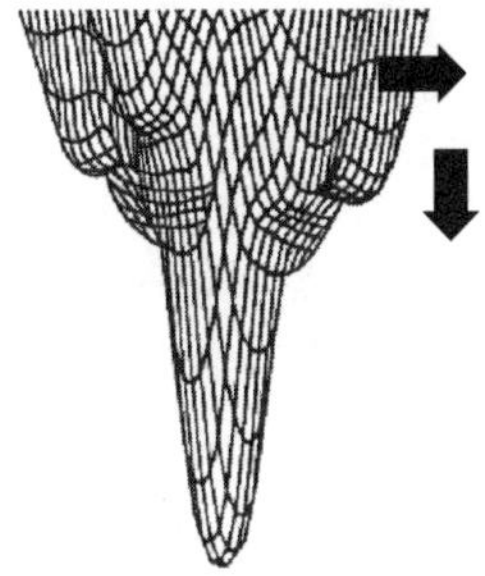

6" wide
20" deep

Foam. Tools for foam application to wall voids can be used to apply foam beneath concrete slabs. Holes drilled along the edges of concrete slabs for liquid injection can be used for foam application. Directional tips are not necessary.

At least 75% of the total treatment must be as liquid, and the remaining 25% or less as foam. The amount of termiticide applied as foam or liquid must be equal to the total amount required by the label.

APPLICATION METHODS

Excavated Soil or Backfill Treatment. This method is used when the soil to be treated is close to open or flowing water, a below-ground well, or water storage. The objective is to prevent termiticide from moving or being carried away from the treatment site and contacting these water sources. The steps, including digging and removing the soil to be treated, then returning the termiticide-treated soil to its original location, are:

1. Measure the length and depth of the treatment site to determine the required volume of insecticide.
2. Trench and remove the excavated soil and place it on or in a waterproof container.
3. Apply liquid termiticide to the soil according to product label directions: 4 gallons per 10 linear feet per foot of depth of the trench.
4. Mix the liquid and soil thoroughly.
5. After the soil has absorbed the termiticide, return the soil to the trench.

Foam. Termiticide foam application is suitable for porches, chimney bases, around and into foundation piers and hollow block voids, behind masonry and other veneers, under concrete slabs, and in wall voids where termites are present. Applications can be a combination of liquid and foam or only foam.

Around plumbing entries, under concrete slabs near cracks, and in areas where the foundation footing is not deep, foam only may be used. Where both lateral spread and vertical penetration is needed to reach the footer, both liquid and foam may be used.

Trenching. For treatment made after the final grading around the foundation, application is made by trenching and rodding into the trench or simply trenching along the foundation walls when the footing is 1 to 2 feet below the surface. The objective of excavating a trench and rodding is to treat soil adjacent to the foundation and the footing. The trench should not extend below the bottom of the footing, and rodding need not extend farther than 4 feet from the soil surface.

The trench should be at least 6 inches deep, about 6 inches wide at the top, and angled away from the foundation to force liquid toward the wall. Soil from the trench must be treated and returned.

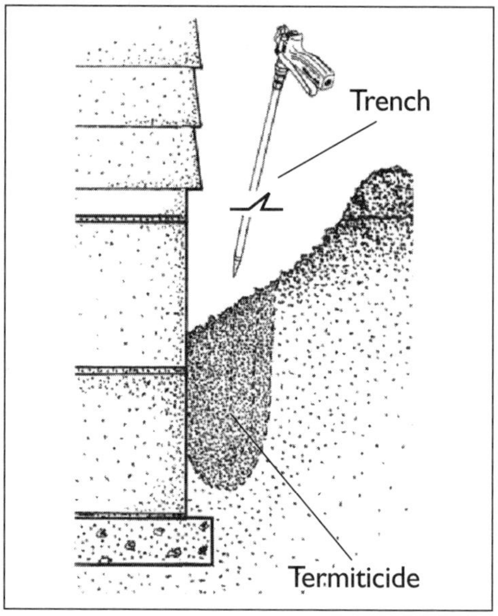

Rodding. Applying liquid termiticide adjacent to the foundation can be done by inserting rods into the soil. Rodding may be in a trench dug along the wall or directly through the soil surface. The depth of the treatment depends on the depth of the footing, but it need not be more than 4 feet.

Termiticide may be released through the rod as it is pushed into the soil to aid in penetration of compacted soil. Soil along the wall is more pervious than soil farther way, and the liquid flow will be toward the foundation wall.

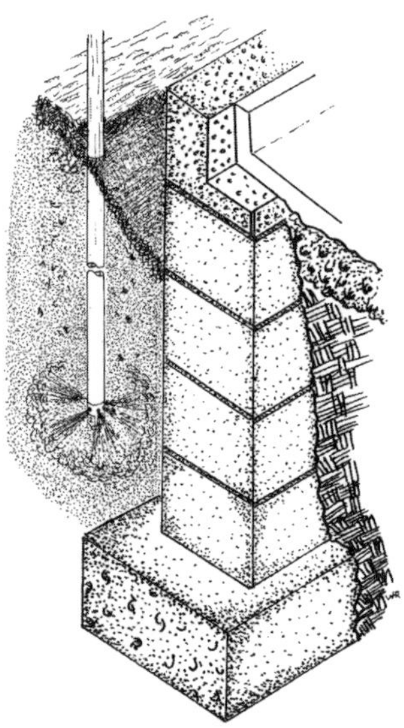

Soil rodding

When loose, sandy soil or other conditions make trenching difficult, rodding may be used without trenching.

Volume Reduction. When termiticide is applied to clay soils, penetration may be difficult because of the limited movement in this soil type. Label directions typically state that if soil will not accept the label application volume, the volume may be reduced, provided there is an increase in concentration so the amount of active ingredient applied to the soil remains the same.

Example: If 0.06% dilution of a termiticide can not be applied to achieve 4 gallons of finished dilution per 10 linear feet per foot of depth, then 0.125% can be substituted and applied at a rate of 2 gallons per 10 linear feet.

When liquid volume is reduced, the soil rodding spacing may be adjusted to account for the lower volume dispersed in the soil. It may be necessary to space soil rod injections closer than 12 inches to create a continuous treated zone.

Wood Injection. Galleries below the wood surface can be drilled or punctured with an injection needle and treated with liquid or foam insecticide. Drywood termite galleries are empty and liquid or foam can travel away from the injection point.

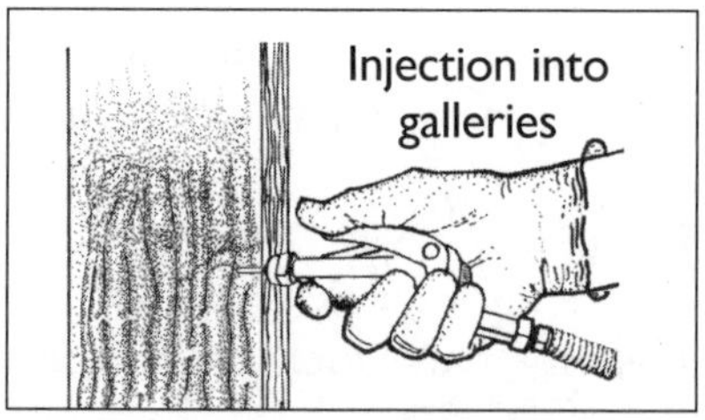

PRE-CONSTRUCTION TREATMENT

Treatments made during all stages of building construction are included in the pre-construction section of termiticide product labels. Application during construction can result in establishing continuous horizontal and vertical zones of treated soil around foundation walls. Treatment at this time can be done before or after backfill soil is in place and final grading completed.

When the top of the foundation footing is exposed or close to the surface, the soil adjacent to the footing must be treated.

Treatments made after backfilling and completion of the final grading must be made by trenching and rodding into the trench to, but not below, the footing.

Soil Surface. Termiticide must be applied to the entire soil surface that is to be covered by a concrete slab. This includes the slab under the living area, carports, porches, stairs,

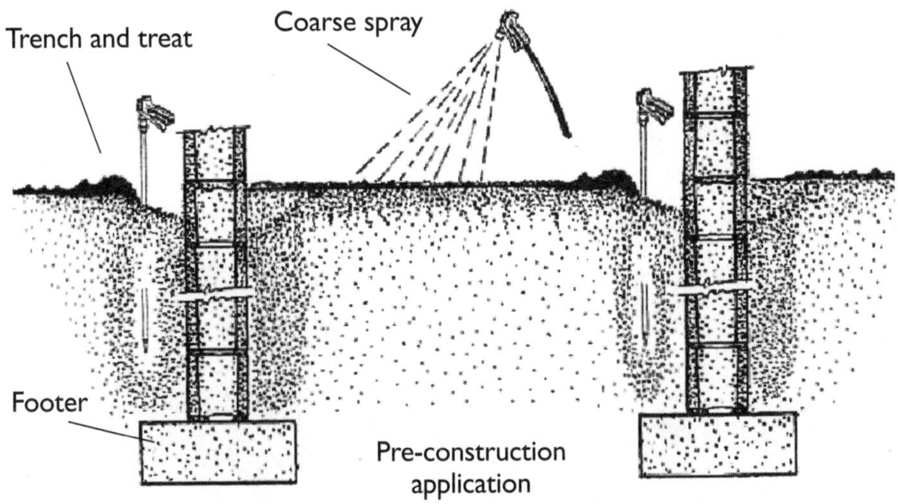

basement floors, and extended entrances. The soil expected to be beneath a crawlspace must be treated. Application should be with a coarse spray of 25- to 50-psi pressure, and with sufficient spray overlap to provide uniform coverage.

- If the fill or soil surface is covered with gravel, a sufficient amount of termiticide must be applied to reach the soil beneath. Label directions may permit increasing application from 1 to 1.5 gallons per 10 square feet.

- If the concrete slab is not poured the same day the soil is treated, all treated soil must be covered with a waterproof sheet to prevent run-off following rainfall. If foundation walls are in place around the foundation soil, the covering is not necessary.

Monolithic Slab. This single-pour concrete slab combines the foundation with the footer and foundation wall. It is of one-piece construction and typically poured directly on the ground. The soil beneath is best treated prior to the pouring of concrete.

Crawlspaces. Foundation areas that will become a crawlspace after construction are accessible during construction. Trenching or trenching and rodding to the top of the footing can treat soil adjacent to walls. When the top of the footing is exposed, the adjacent soil must be treated.

Hollow-Block Foundation Walls. The top of block foundation walls may be accessible

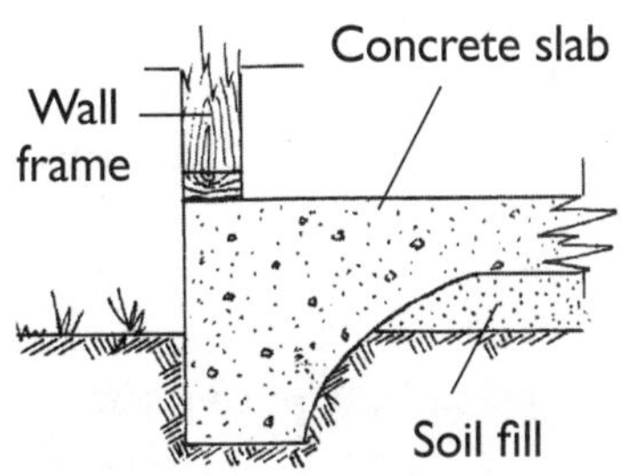

Monolithic slab

when concrete slabs are poured. In the case of supported slabs; the concrete may be poured to the outer edge of the block, thereby sealing the top block, or only to the center of the block, leaving it accessible for treatment.

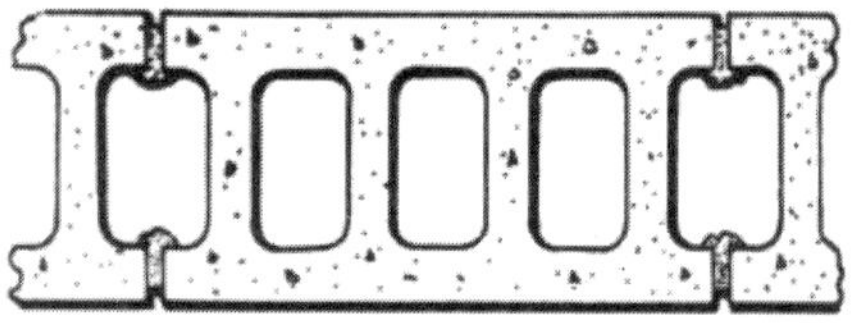

Masonry block

Voids of masonry blocks may be drilled and treated if the top of the block wall is not open and accessible. Access holes can be drilled as close as possible to the footing.

POST-CONSTRUCTION TREATMENT METHODS

The post-construction time frame is considered to be after the soil grading is done and the building construction is completed or nearly so.

Post-construction treatment guidelines on product labels are prepared for monolithic, floating, and supported concrete slabs.

Exterior Perimeter. The exterior soil is treated by trenching and rodding. Because the backfill and final grading have been done, the actual depth of the treatment will depend on the depth of the footing.

- *Garage, driveway.* The concrete slab of attached garages should be included in the exterior perimeter treatment plan. Sub-slab application can be made from inside the garage or by horizontal rodding from outside. Treat along interior partition walls and walls adjacent to the main structure.

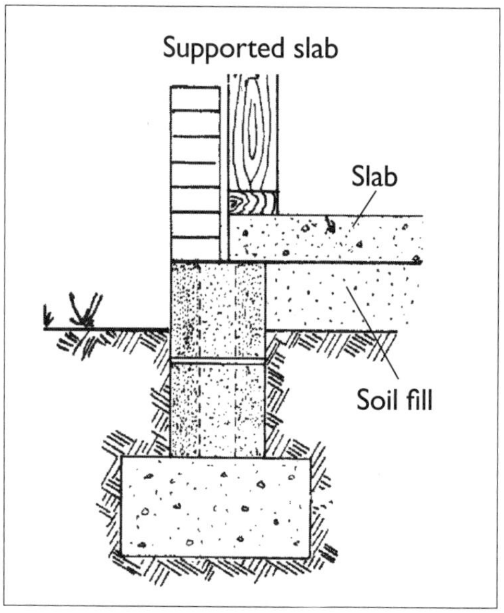

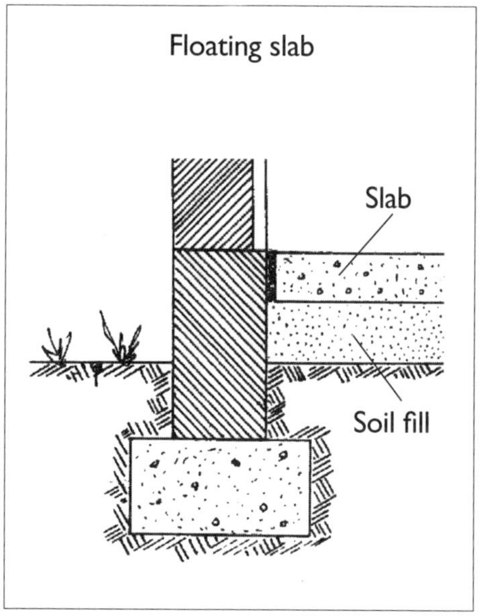

When the footing is more than 4 feet below the soil surface, application is made by trenching and rodding into the trench to a depth of 4 feet. The ultimate depth achieved by the liquid termiticide will depend on soil type and degree of compaction around the foundation.

Garages attached to the house are often the initial infestation site for subterranean termites.

Driveways may be attached to the concrete slab of the garage. Treatment should be made around the building supports or wall elements that are located at the driveway junction with the garage slab. Drill and rod the soil at 1-foot intervals; use a directional nozzle to guide the liquid toward the junction.

- *Window wells.* Basement window wells are usually recessed into the ground. They can be built with concrete, concrete block, brick, or galvanized metal. At the bottom may be concrete or a layer of gravel for drainage. Treatment of window wells includes trenching and rodding around the perimeter and rodding inside. If the floor is concrete, it must be drilled and rodded.

Interior Perimeter, Sub-Slab Injection. The soil under concrete floor slabs is treated by drilling vertically through the slab along the interior perimeter of the foundation. Holes must be drilled along concrete expansion joints, cracks, plumbing, and utility services penetrating the slab. A directional tip will direct the termiticide to the foundation wall. Drill holes should not be spaced more than 1 foot apart to achieve a continuous treatment zone in the soil adjacent to the foundation.

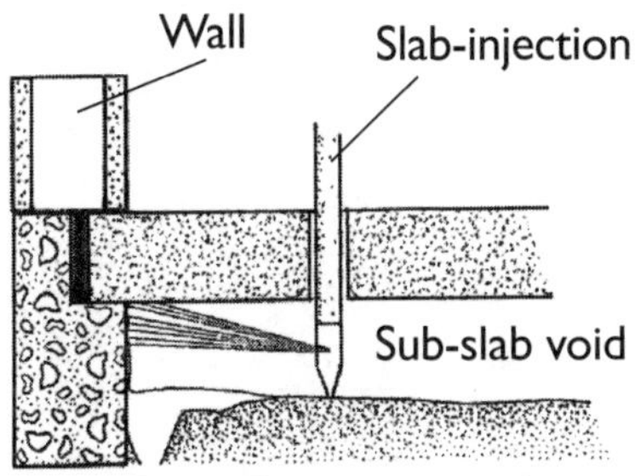

Interior perimeter

- *Vertical drilling.* Sub-slab treatments are made from inside the structure. Application is made by drilling holes through the slab at intervals around the perimeter.

  Treatment holes may be drilled adjacent to partition walls where there is evidence of termite activity in the wall or in the soil beneath the wall. A sub-slab injector can be used when interior floors are finished (e.g., in tile or wood), and a 180° or 360° directional tip should be used to ensure uniform application of the liquid termiticide on the soil surface.

  Spacing of drill holes is critical to the formation of a continuous treatment zone beneath concrete slabs. Drill holes should be spaced 1 foot apart and a directional tip used.

  Using a straight tip or spacing the drill holes more than 1 foot apart may leave an untreated gap in soil coverage and permit termites to enter.

- *Bath traps.* Soil around plumbing and drain pipes in bathrooms must be treated. If access to the space beneath bathrooms is limited, a small door or inspection vent should be installed. Treatment around bath traps and shower drains is done with vertical rodding or drenching of the soil surface.

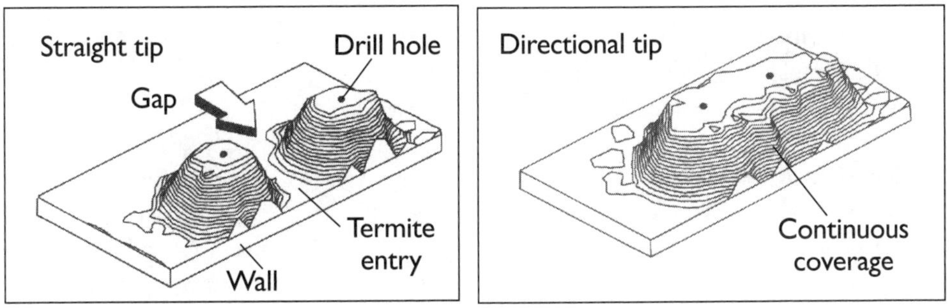

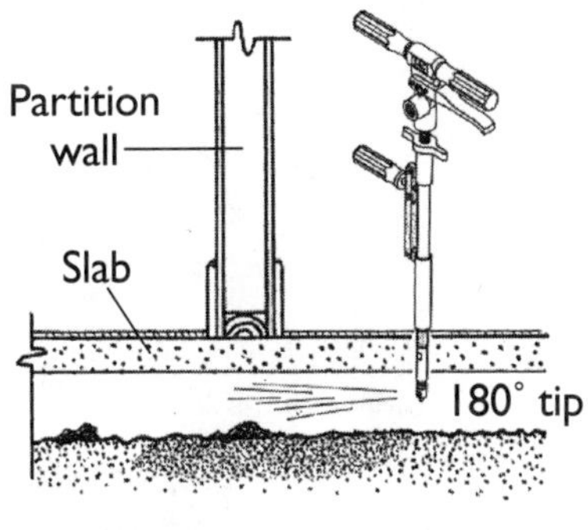

Sub-slab treatment

- *Horizontal drilling.* Treating beneath concrete slabs from the outside is necessary when floor coverings or interior design features make vertical drilling impractical or unwanted. Application can be made by drilling horizontal holes in the exterior walls and rodding a short distance into the soil beneath the slab. This method is usually called "short rodding" on termiticide product labels.

Crawlspace Construction. Crawlspaces can be treated by standard trenching and

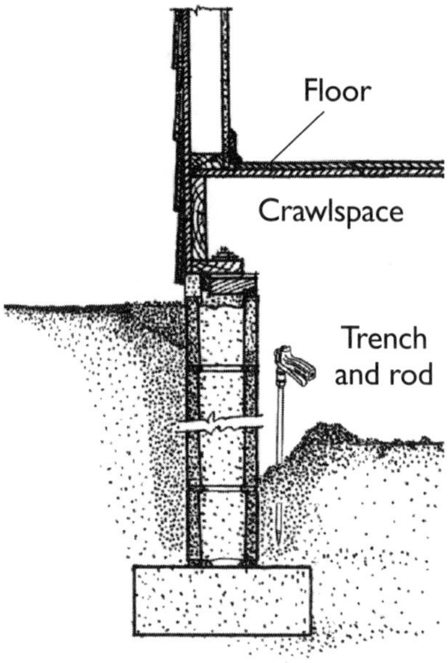

Crawlspace treatment

rodding adjacent to the foundation wall. Inaccessible crawlspaces must be treated by soil drenching, drilling through the floor above, or horizontal rodding of the foundation wall.

- *Vapor barrier.* Soil in crawlspaces may be partially covered with a plastic vapor barrier against soil moisture. Vapor barriers must be rolled back to treat the soil, and the plastic sheet must not be punctured. After treatment, replace all sheets in their original position. If the vapor barrier is composed of a thin layer of concrete, this must be drilled and the underlying soil treated.

TERMITICIDE RE-TREATMENT

Re-treatment for subterranean termites can only be performed if there is evidence of re-infestation or disruption of the original treated zone.

Re-treatment is permitted when there is a breakdown of termiticide in the original treatment zone. It is conducted with the standard methods of trenching and rodding.

LOCALIZED INTERIOR TREATMENT (EPLIT)

Exterior Perimeter/Localized Interior Treatment (EPLIT) makes interior treatment optional if there is no termite activity inside.

The interior receives treatment only in the areas that have termite activity. If no termite activity is observed inside the structure, interior local treatments are not required. The criteria for treating are: presence of winged termites, live termites, exposed shelter tubes, or live termites in wood. Interior application may be made around plumbing and utility lines that penetrate floors, around bath traps, and along concrete expansion joints and cracks. If termite activity exists in occupied areas, or unoccupied areas such as crawlspaces, a localized treatment must be made at least 2 feet in two or more directions from the site of activity.

Crawlspace. When termite activity is found within a crawlspace, the location of the infestation must be treated. In addition, a localized interior treatment must be made at least 2 feet in either direction from the termite activity.

Interior Walls. Treatment of interior walls includes application of liquid or foam to the area under the floor and behind the wall adjacent to the damage. Liquid or foam should be used to treat a 2-foot area around the infested site.

Hollow-Block Foundation. When termite activity occurs in or within 2 feet of a hollow block wall or void in masonry, the wall adjacent to the damage must be drilled and treated with liquid or foam. Application must be in a 2-foot area around the termite activity. Access holes should be drilled as close to the footing as practical. Drilled holes must be plugged after treatment.

Bath Traps, Shower Drains. When termite activity is found within a 2-foot radius of the bath trap, the soil around the plumbing and drainpipe entry must be treated. The soil in this area may be covered with concrete or other material; this must be removed before rodding or soil drenching with liquid or foam. When termite activity is found under the shower drain, the concrete slab must be drilled and the underlying soil treated, using a directional nozzle or foam to direct termiticide to the 2-foot treatment zone. Horizontal rodding may be used to access and treat the soil beneath the shower drain or bath trap.

BAITING SYSTEMS

Termite baiting systems are designed to induce foraging workers from a colony to include one or more cellulose food sources into their set of satellite feeding sites. Once workers accept the new food source, they are exposed to an insecticide that will ultimately disrupt or kill the colony.

Foraging. The efficacy of baiting systems relies on the foraging activity of workers and their ability to feed on or become exposed to the toxic material and share it with nestmates within a certain period of time. The acceptance of bait stations as feeding sites by foraging workers is often unpredictable. In spite of one or more termite colonies in the vicinity, there may be little or no feeding activity at stations for weeks or months.

Seasonal Conditions. The ability of termites to locate bait stations varies, based on foraging intensity, time of year, and temperature and humidity. Foraging activity typically increases during late winter and spring and declines in fall and winter. Consumption of the cellulose matrix in bait stations is closely linked to ground temperature and moisture. Feeding peaks in summer when ground temperatures are 72°F to 75°F and humidity is above 80%.

In general, termites do not favor bait stations that are in the open and exposed to sunlight and temperature extremes. These stations may have little or no activity in summer and winter. At bait stations in shaded locations, termite feeding usually peaks during summer.

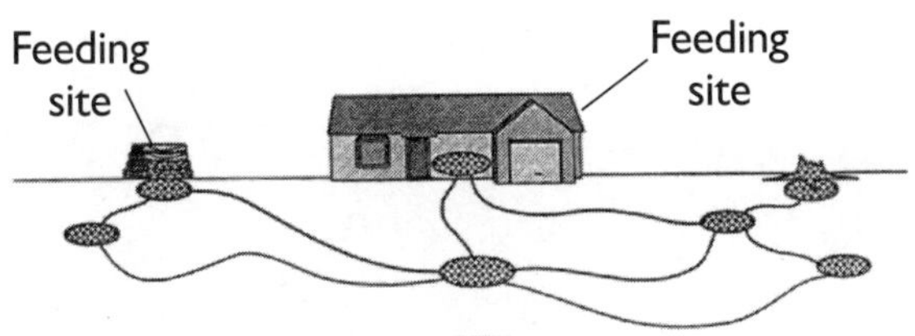

Subterranean termite colony with feeding sites outside and inside a structure

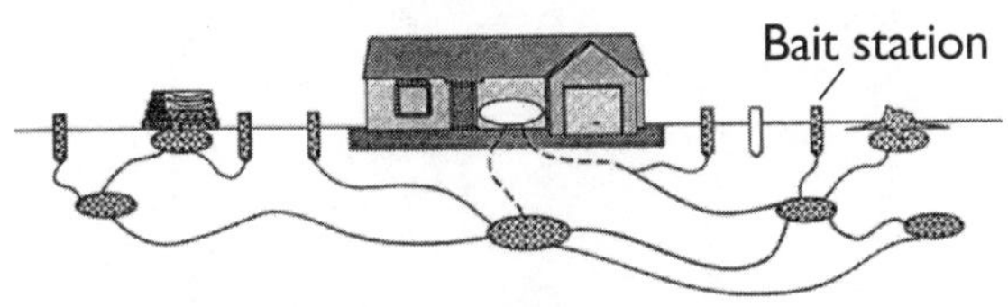

Termiticide applied to the foundation can eliminate indoor infestation, and baits provide additional feeding sites

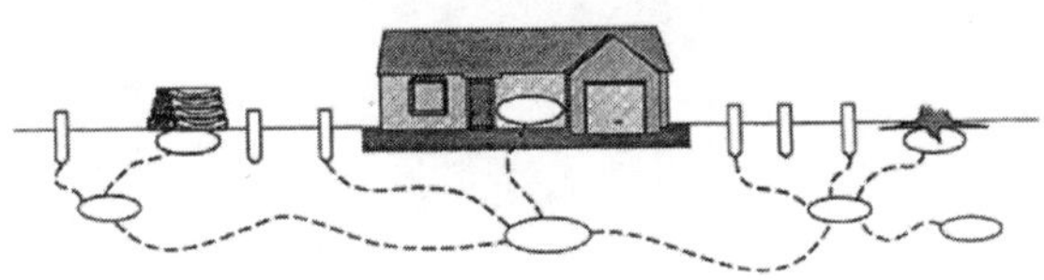

Baits can eliminate satellite feeding sites

CHAPTER 12

WOOD-INFESTING INSECT CONTROL

Ants, bees, and beetles infest natural and structural wood in all habitats. Carpenter ants and carpenter bees live in colonies, and they build nests in exposed or moisture-damaged wood. Wood-infesting beetles are active throughout the year because the wood they infest is the food source for their larval stages. The wood infested may be hardwood or softwood and is usually not moisture damaged. Continued infestation by some species can result in the loss of structural integrity of the wood. Sources of these pests are natural populations in empty lots or wooded areas, and infested structures scattered in urban and suburban areas.

CARPENTER ANTS

In many regions, carpenter ants (*Camponotus*) are the most important wood-infesting insects in the urban environment. They are called carpenter ants because nest galleries have a smooth surface, as if sanded by a carpenter. Carpenter ants do not eat wood. They are outdoor scavengers and usually feed on insects and aphid honeydew. However, workers will forage indoors and are often found in kitchens searching for food. The ants are a threat to structural timber because of their ability to excavate large nests and the potential of continuously infested wood being weakened.

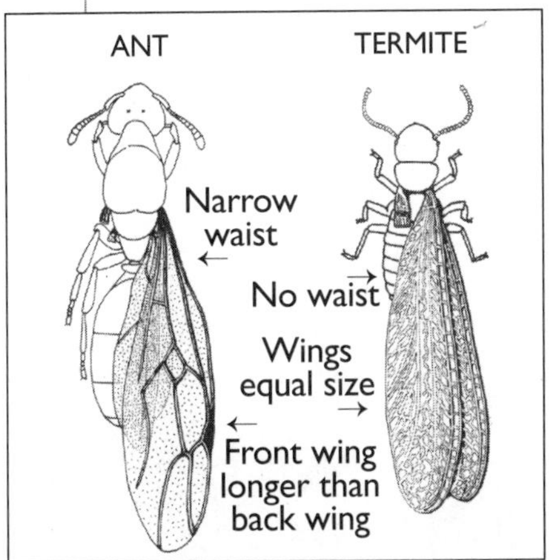

Identification and Distribution. Carpenter ants are large; they may be all black, black and red, or reddish brown. The head and body of major workers is much larger than those of the small, minor workers. The wings of both the male and female are of different sizes, distinguishing them from adult termites, in which the wings are the same size and length. Colonies produce winged adults in the spring and summer; swarms usually occur in late afternoon on warm, sunny, humid days.

Colony Activity. Colonies usually have 8,000 to 10,000 individuals and are divided into small, satellite nests and a main or parent nest. The parent nest is usually in a humid environment and contains the queen, developing eggs, and small larvae. In the satellites, there are workers, larvae, pupae, and sometimes winged reproductives. The number of satellite nests is variable and depends on environmental conditions and colony age.

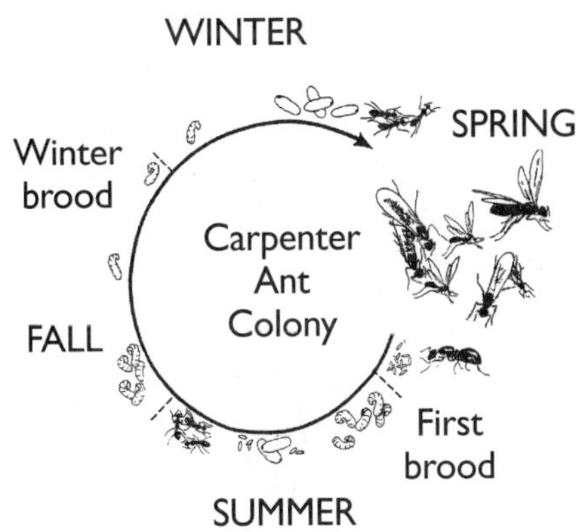

Food. Natural food includes honeydew, plant sap, and live and dead insects. Indoors, the ants forage for sweets and high-protein foods. During spring and early summer, the workers prefer proteins, which are fed to developing larvae. In late summer and fall, workers prefer carbohydrates, which are used as an energy source. The majority of foraging is done at night, and most species rely on chemical trails for orientation to a food source. Indoors, they travel on edges formed along walls and cabinets, and they orient to pathways formed by electrical wires and water pipes. On the outside of structures, they often travel along the branches of trees that contact the roof or siding.

Infestation Sites. Nests are constructed in live or dead trees, in rotting logs or stumps, or in sound or moisture-damaged structural wood. Some species nest in soil, under rock, or in logs, such as firewood. Workers excavate the wood with their large toothed mandibles. As workers age, the teeth on their mandibles wear down. Galleries are irregular in shape.

Chewing sounds made by excavating workers, and rustling sounds of workers moving in galleries are audible. When disturbed, Pacific Northwest *(Camponotus modoc)* workers often strike the walls of the tunnels with their mandibles.

When the nest is indoors, the queen and developing larvae are usually located where there is a constant source of water. This may be from a pipe leak or poor drainage in a bathroom or kitchen.

Satellite Nests. Indoor infestations are often satellite nests, which are connected to a main colony located outdoors. Satellite nests may remain unnoticed because workers forage outdoors and in the evening. Nest locations include: in wall voids, roof supports, sub-flooring, ceiling joists, and sill plates or foundation piers.

Frass. The fibrous pieces of wood from carpenter ant nests are concentrated in unused galleries or mixed with other colony debris and pushed out of the nest through small openings. The pieces are about 1/4 inch long. Tunneling in sound wood produces frass uniform in color; but frass from moisture-damaged wood may be a mix of dark and light pieces. Carpenter-ant frass usually contains small pieces of soil and gravel, and pieces of uneaten parts of dead ants and other insects.

Pavement ants often establish nests between foundation walls and the wood sill plate of houses. Excavating the nest produces frass similar to that of carpenter ants. It is usually composed of small seeds, fragments of dead insects, and fibrous wood pieces. The frass is usually located directly at the nest site.

ACROBAT ANTS

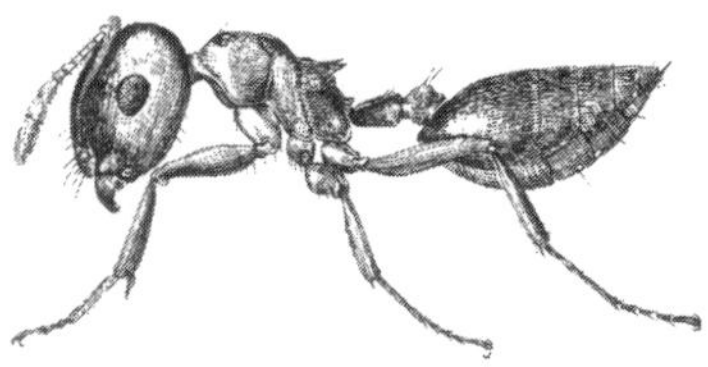

Acrobat ant

Acrobat ants (*Crematogaster*) build nests above ground in tree stumps and in moisture-damaged structural wood. They are small, and infestations often remain unnoticed until workers are observed or frass is seen falling from infested timbers. Workers are distinguished by their broad head and heart-shaped abdomen (gaster). They usually have a reddish brown head and dark brown to black gaster.

Colonies contain 2,000 to 3,000 workers. The common name, acrobat ant, is derived from their habit of raising their gaster over their head and thorax when disturbed. Natural food includes live and dead insects and honeydew. Indoors, they prefer sweets and meat. Swarmers emerge from June to November. Workers in large colonies can be aggressive and, when disturbed, emit an objectionable odor.

Frass. The frass particles of the acrobat ant are small and fibrous. They are easily carried away by the wind and may not be noticed until the nest is large.

CARPENTER BEES

Carpenter bee

Carpenter bees (*Xylocopa*) are large, with yellow and black patterns; they resemble bumble bees. They create tunnels in wood, including seasoned hardwoods, softwoods, and decaying wood. Females are not usually aggressive, but are capable of stinging. Males are aggressive and threatening when they guard nest sites, but they are not able to sting.

Colony Activity. Mating occurs in early spring following territorial flights by the males. Females may use an old gallery, create a new one, or make a new one from an entrance

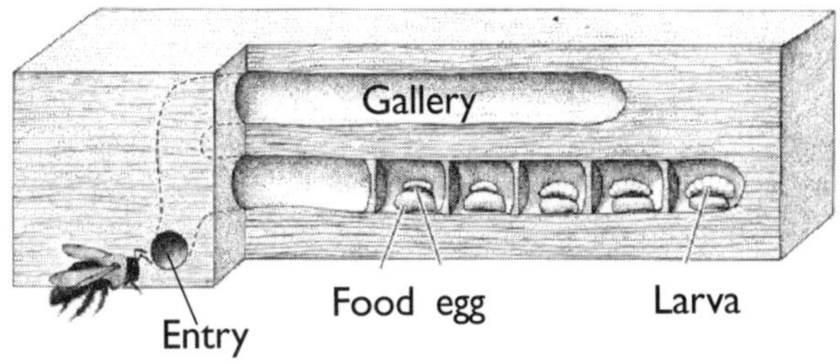

used by several other females. Adults that have over-wintered die during the summer of the second year, after they have mated and produced galleries and eggs for the next generation. There is usually one generation per year.

Females provision each cell in the gallery with a small amount of pollen and nectar, then lay an egg on it. This cell is sealed with chewed wood pulp. This procedure of provisioning, egg-laying, and sealing is repeated until a series of about six cells is formed in the gallery. Larval and pupal development time is about 36 days. The first bee to become an adult is at the end of the gallery and has to cut through the partitions of all the cells ahead of it to emerge from the nest. The other adults do the same. For the remainder of the season, adult bees gather and store food in the galleries for use during inclement weather. During this period, there is no mating or nest construction.

Infestation Sites. Exposed structural timbers on decks, fences, and siding can be continuously infested, and the wood may become structurally unsound. Female bees gnaw into the wood surface to make a hole about 1/2 inches in diameter; they can tunnel several inches per day. Galleries in new nest sites extend about 12 inches. Galleries may be used by several generations of bees, and the system of tunnels can become very elaborate. More than one female may use the same entry hole, but their brood tunnels will be separate.

Frass. The galleries of carpenter bees are empty of frass. The pieces of wood shavings resulting from the female tunneling in the wood are scattered below the nest opening. The frass may accumulate in undisturbed sites and is a good indicator of carpenter bee activity. The pieces are 1/4 inch long, fibrous, and irregularly shaped. The composition is only wood shavings, unlike the frass of carpenter ants, which includes fragments of dead ants and soil.

Carpenter bee frass

POWDERPOST BEETLES

Powderpost beetle is a general name applied to species in the families Anobiidae and Lyctidae. The name comes from the ability of the larval stage to feed and reduce wood to a powder. There are several Anobiid powderpost beetles that attack structural and ornamental wood. Most species have similar life cycles—including egg, larval, and pupal development.

Anobiidae. Females prefer to lay eggs on rough-surfaced wood that is 2 to 5 years old. First-stage larvae bore in the direction of the wood grain. Galleries are packed with fecal pellet frass. Larvae stop or reduce feeding in response to low winter temperatures, and resume feeding in spring. Full-grown larvae tunnel close to the surface and prepare a pupal chamber. Adults cut the exit holes in the wood surface, then begin emerging in spring and continue through mid-summer. Adults do not feed and are nocturnal. The life cycle is 1 to 5 years, depending on the quality of wood infested and fluctuations in temperature and wood moisture.

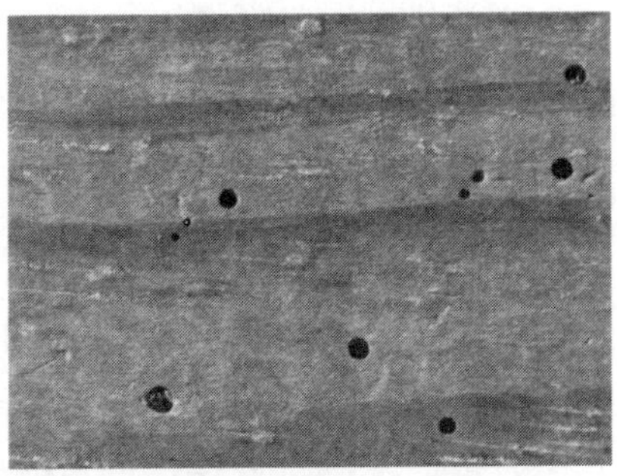

Emergence holes

Emergence holes of wasp parasites of anobiid beetle larvae are often seen in wood that has been infested for a long time; the wasp emergence holes are about 1/3 the size of the beetle emergence holes.

Infestation Sites. These beetles infest both hardwood and softwood lumber, but they are most often found in softwoods. Exit holes are round and about 1/8 inch in diameter. Anobiid beetles occur outdoors in dead wood; they enter houses from outdoor infestations and spread to unfinished wood indoors. While there is generally little unfinished wood in living areas, there is plenty in basements and crawlspaces. Temperature, humidity, and the condition of wood limit larval development. Adults produced from wood in dry environments are generally small. These adults produce fewer eggs, and, therefore, infestations decrease. This may explain why infestations decline when wood moisture levels remain low.

Anobiid larval development is influenced by temperature and moisture content of the infested wood. Crawlspace ventilation of 1.0-square-foot opening per 25 linear feet of wall can reduce wood moisture content. The use of 4.0- to 6.0-mil thick plastic sheets to cover exposed soil can reduce the movement of moisture from soil to exposed wood.

Reducing wood moisture in structural timber is a slow process, and any influence it would have on anobiid beetle infestations would be equally slow and not practical as a control strategy.

Frass. Anobiid beetle frass has a gritty texture; the grit is actually small pieces of tightly compressed frass. Anobiid frass remains somewhat loose and is easily dislodged from galleries.

Old and inactive infestations of anobiids in crawlspaces and basements can give the false impression that they are active, because frass is jarred loose from emergence holes by activity in the living space above.

Lyctid beetle exit holes

Lyctidae. Female beetles chew a small amount of wood to assure the starch content is above 3% before laying eggs. Eggs are placed in exposed pores of the wood; oak and ash have large pores and, therefore, are often infested. The larval stage lasts 2 to 9 months, but development may be extended to 2 or more years in old wood. Full grown larvae tunnel close to the wood surface and form a pupal chamber about 1/8 inch below. The adult cuts the exit hole to emerge. These beetles can fly and are attracted to light, including ultraviolet light traps.

Infestation Sites. These beetle larvae limit their feeding to hardwoods. The round exit holes made by the adults are slightly less than 1/8 inch in diameter. Lyctid beetles feed in air- or kiln-dried hardwoods; wood less than 5 years old is preferred because the starch content is high. Infestations occur in hardwood flooring, furniture, and picture frames.

In air-dried hardwood lumber much of the starch and other carbohydrates are naturally eliminated. In kiln-dried wood there is little or no loss of natural starch in the wood. This wood is susceptible to attack for many years.

Bamboo is often infested with lyctids because of its high starch content.

Frass. The frass of lyctid beetles is soft and powdery, with no gritty feel. Frass from lyctid beetles is packed into the galleries behind the larva as it feeds and is not easily knocked from the galleries.

AMBROSIA BEETLES

Ambrosia beetle adults bore into live and recently felled trees. The adult beetles and larval stages do not eat wood, but instead feed on a fungus, called ambrosia, that grows in the galleries the adult beetles excavate. As the fungus grows in the galleries, it stains the walls dark blue to black. The galleries are not extensive and when larval development is complete, the adults exit the wood through the holes made by their parents. The emergence holes are about 1/16 inch in diameter and the perimeter is stained blue or black. There is no frass produced by ambrosia beetle larvae.

The damage caused by ambrosia beetles is cosmetic; there is no structural weakening of the wood. The only problem is caused by the dark staining of the few galleries that might be visible at the surface. Ambrosia beetle damage is often seen on the logs of modern log houses, and on lumber for traditional frame houses. They do not re-infest structural wood because the fungi require high wood moisture to survive.

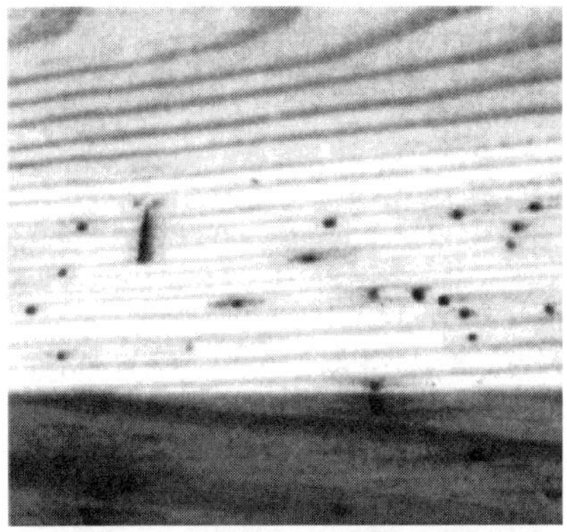

Ambrosia beetle damage

LONGHORNED BEETLES

Longhorned beetle larvae feed in the wood of dead and downed trees, infesting both hardwood and softwood trees. A few species attack pine trees soon after felling and can be pests in modern log houses.

***Old House Borer** (Hylotrupes bajulus).* Adults lay eggs in the surface of seasoned wood. Hatching occurs in about 9 days and is not prolonged because of environmental conditions. First-stage larvae bore immediately into the wood and begin feeding. The sound made by the mandibles of medium and large larvae can be heard at a distance. Full-grown larvae tunnel to the surface and cut an oval exit hole; the pupal chamber is below the exit hole. Development time in structural wood is 3 to 5 years, but can extend to 10 years or more. Adults emerge in June and July and live about 15 days.

Larvae of the old house borer feed in seasoned pine, spruce, and fir. The larvae have an enlarged head and thoracic region; the remainder of the body tapers to the end. Larvae feed in tunnels packed with powdery or fibrous frass. Adults have long antennae or "horns"; they are active during the day, but do not feed and live only a few weeks.

The old house borer does not occur naturally in dead trees or firewood. It re-infests seasoned lumber, and the life cycle can extend for 5 or more years. Wood moisture of less than 10% and temperature fluctuations prolong development but do not kill the larvae.

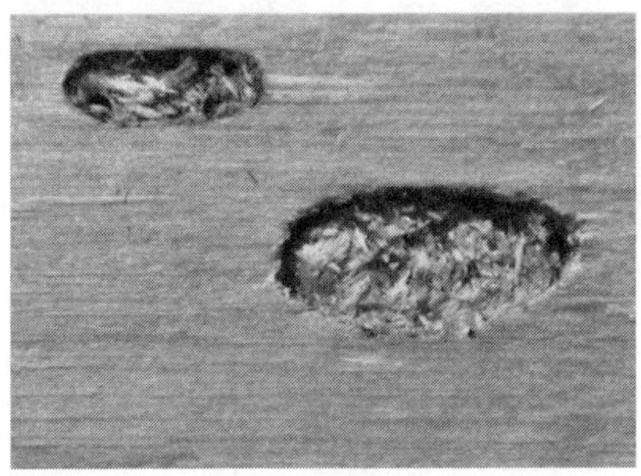

Pine sawyer galleries

Pine sawyers attack recently cut logs, which have a moisture content high enough for them to complete development. Damage to wood from sawyers is often misidentified with damage caused by the old house borer.

***Pine sawyer** (Monochamus spp.).* Eggs are laid in the bark of trees or in bark that remains on recently felled logs. Early-stage larvae feed beneath the bark, late-stage larvae feed deep in the wood. Development is completed in 1 or 2 years, and adults emerge in April and May. Larval tunnels are oval or round, and filled with fibrous frass. Full-grown larvae tunnel to the surface and make the exit hole. The pupal chamber is below the frass-plugged exit hole; the adult pushes through the plug to emerge. Larvae are called sawyers because of the sawing sound they make while feeding. Sawyers do not re-infest the original wood.

Infestation Sites

- *Wood-frame houses.* Old house borer infestations occur primarily in houses that are less than 10 years old. Adults and larvae can occur in lumber stored outdoors in lumberyards, but not from natural habitats. Development time of larvae in different household locations is different, with the attic being the most delayed because of the wood moisture content and temperature fluctuation.

- *Log houses.* The old house borer infests logs and other raw timber used in modern log houses; infestations originate at the manufacturing site while logs are held for processing. Infestation at the construction site is unlikely. Adult beetles do not come from surrounding wooded areas to infest the logs, because there are no natural populations of this beetle. The old house borer is not native to the U.S., it originated outside North America. The high moisture content of the logs favors larval development, and the life cycle can be completed in 2 or 3 years. Re-infestation of logs occurs from adults emerging indoors and outdoors, and females remaining close to the infestation site to lay eggs. Development time for succeeding generations is usually more than 3 years.

Old house borer frass

Frass. Old house borer frass is composed of barrel-shaped pellets of digested wood and irregular shaped particles. It is tightly packed into the tunnels behind the larva as it feeds, and it appears as a fine powder when tunnels are opened. A small amount of powdery frass is pushed from the emergence hole when the adults emerge.

CONTROL STRATEGIES

Control and prevention of carpenter ants, bees, and wood-infesting beetles is based primarily on insecticide application to the wood surface or injection below the surface. While fumigation can be effective, it is rarely necessary or economical, because infestations are usually limited to distinct sites.

Modern insecticides, such as pyrethroids, provide quick knockdown and residual activity to kill these pests. Professional application equipment provides precision application of dust, liquid, foam, or bait.

Chemical control must be based on clear evidence of an active infestation. Distinguishing an active infestation of wood-infesting beetles and carpenter ants is problematic. It should not be based solely on the presence of a few worker ants found indoors, or pieces of wood with beetle emergence holes and falling frass. Carpenter ants have a wide foraging territory and can move far from the nest site. Beetle emergence holes will remain unchanged for years, and frass can drop out of the hole long after the initial infestation has ended.

Carpenter ants are selective feeders, with seasonal food preferences. Colonies often exhibit a shift in food preference following the swarming of reproductives. Prior to over-wintering, carbohydrate-rich foods are often preferred by *Camponotus* species.

Baits: Carpenter Ants. Ant control baits are useful when the primary nest site cannot be found, or where location or customer preference limits the use of other insecticide formulations. Baits are generally slow acting, but are effective because they eventually kill the queen. They are most effective when eaten by foraging workers and transferred back to the colony. Baits require careful placement to intercept trails of foraging workers. Without access to a large number of foraging workers, a lethal amount of insecticide will not be carried back to the nest.

- *Application.* Position bait and/or liquid bait stations to maximize access for foraging worker ants. The efficacy of baits depends on the distribution so that a large number of colony workers, including those in satellite colonies, have access. Structural infestations that contain large colonies or several sites with developing larvae and workers may not be eliminated if competing sources of food are available. Likewise, workers entering structures from colonies established outside are not likely to carry back enough toxic bait to destroy the colony.

Wall Void Treatment: Carpenter Ants. Open spaces behind walls and cabinets, around door and window frames, beneath floors, and behind roof soffits are common infestation sites. They can be partially filled with insulation and include electric wires, cables, and pipes. In spite of, or perhaps because of, these features, voids provide a protected harborage for primary and satellite nests, especially when there is moisture-damaged wood and outdoor access.

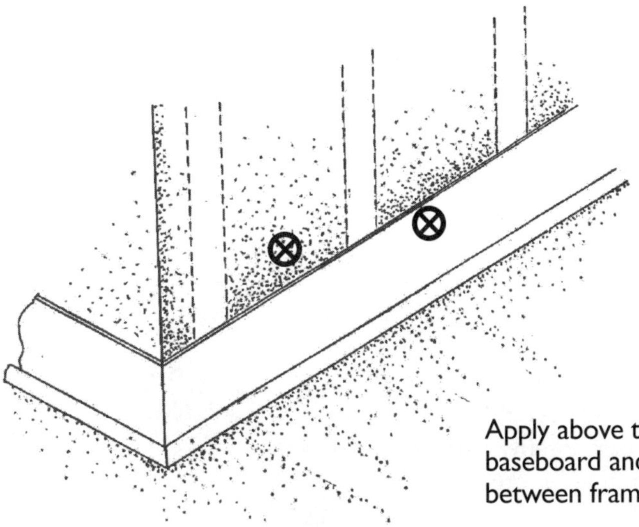

Apply above the baseboard and between framing studs.

Wall Void Treatment

- *Application.* Select a treatment access point that will give the greatest amount of insecticide coverage. In the Pacific Northwest, colonies can have 10 to 15 satellites in locations of the house. When this seems to be the case, treating a large number of the wall voids in a house will eliminate all possible nest sites and provide the best control. In situations where the primary nest site is found and satellite nests are few, treating only a limited number of wall voids may be effective.

- *Foam* will cover exposed surfaces and establish a long-term residue. It is usually applied at the bottom of a wall void, immediately above the baseboard. The treatment point may be at the level of the nest site of carpenter ants. A curved tip will push back insulation, and moving the tip from side to side during application will cover the exposed sill plate and bottom of the framing studs.

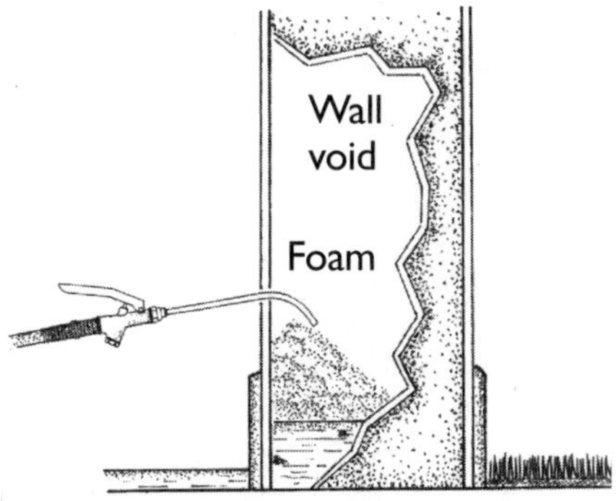

- *Dust* can cover all exposed surfaces and give long-term protection. It can be applied easily with a bulb duster. A pressurized duster can treat a large space, especially voids that do not contain insulation. Application can be made at electrical outlets and cable outlet boxes with a bulb duster and a plastic crack-and-crevice tip. Outlets are close to the bottom of the void, and there is access to the inside if insulation is present.

Caution: To prevent electric shock and severe injury, dust application to electric outlets should only be made with a rubber bulb duster fitted with a plastic tip. A metal tip inserted into an electric outlet will conduct electricity and may cause a short circuit or deliver an electric shock to the applicator.

- *Liquid spray* can be applied to wall voids that do not contain electric outlets or insulation (interior walls). The bottom and sides of the void can be treated with a coarse fan spray, treating sufficiently to wet the surface but not pool at the bottom. The access hole should be large enough to accept the nozzle and the curved end of the extension.

- *Aerosol sprays* will give quick knockdown of pests, but the small droplets have only limited residual activity. Droplets can disperse in an empty space, but are easily blocked if the void contains insulation.

Gallery Injection: Carpenter Ants, Carpenter Bees. Galleries in carpenter ant nests are open and interconnected to form a continuous void. Galleries may extend into relatively sound wood.

- *Application.* Liquid, foam, or dust injected directly into the nest can move through the network of galleries. A special wood injection tip can be pushed through the wood surface to intersect galleries, or the tip can be inserted through a hole drilled into the wood. Using several application sites will result in thorough treatment of the nest. Dust injected into galleries may move only a short distance but can be carried and transferred from worker to worker.

Injection into galleries

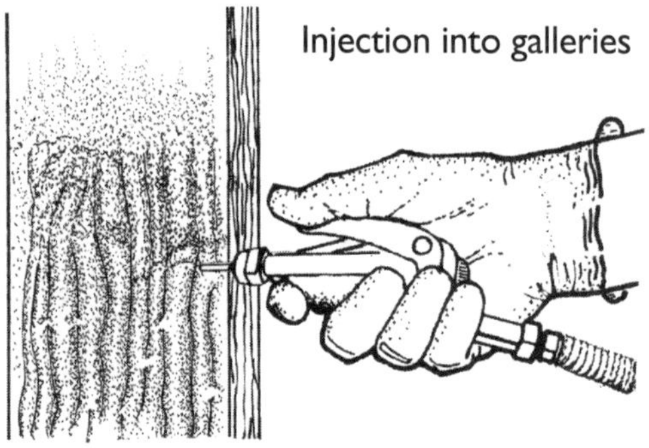

Carpenter ant galleries may not be close to the surface; drilling a small hole into the wood can provide access to several locations and increase distribution of insecticide.

When the entry hole and adjacent tunnel of a carpenter bee nest is treated, the residue can be lethal to all the bees entering or leaving.

Tunnel Injection: Wood-Boring Beetles. The feeding tunnels of wood-infesting beetles can not be treated by injection from the wood surface, or through exit holes. The feeding tunnels of powderpost beetles and old house borers are tightly packed with frass; treating into these galleries will likely result in liquid splash-back. This is not an effective method for treating or controlling wood-infesting beetles; surface or wood injection is more effective and economical.

Pine-sawyer tunnels are open, because the infestation occurred before the wood was processed into lumber. Treating these tunnels is not necessary, since there would be no active infestation of larvae.

Injection into Solid Wood. When liquid is injected 2 to 3 inches below the wood surface, the residue will remain effective for 5 years or more depending on the insecticide. Long-term stability below the surface is due to insecticides binding to organic components of the wood, and having no exposure to environmental conditions. Benefits of injection into sound wood include insecticide placement where the target pest will contact it, increased efficacy and residual activity, and no exposure to people or pets.

- *Chemical fence.* Injecting liquid into the uninfested, solid wood at the edges of an infestation of wood-infesting beetles, carpenter ants, and carpenter bees can stop larval feeding or nest expansion beyond the treatment point. This treatment creates a zone of treated wood—a "chemical fence"—that prevents further activity. The liquid insecticide injected into solid wood disperses on both sides of the application tip.

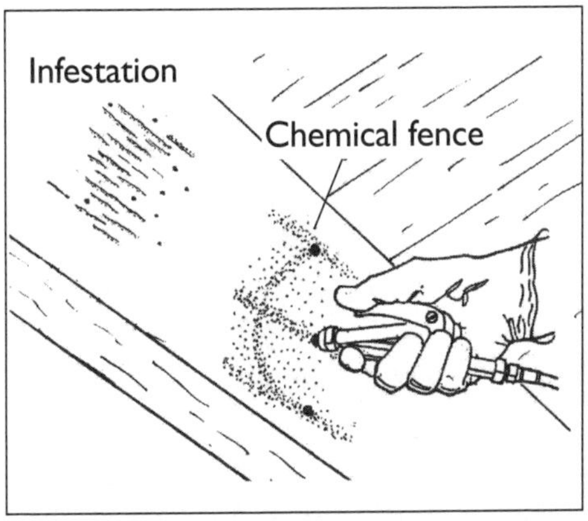

  - o **Movement**. The liquid insecticide injected below the surface will intersect the longitudinal bundles of tracheids. These hollow tubes will carry the liquid in both directions from the injection site, along the grain of the wood. The distance carried is variable, but 6 to 8 inches can be expected. The distance depends somewhat on the wood species. The least amount of longitudinal movement is in spruce, and the greatest movement occurs in pine.

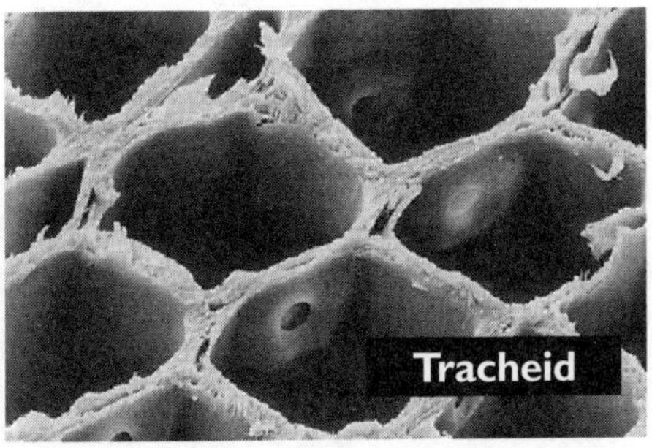

  - o **Application**. Establishing a chemical fence requires drilling about halfway into the thickness of the lumber being treated. The size of the drill bit must correspond to the wood-injection tip; for example, the B&G 1-inch wood-injection tip (Whit-1) requires a 7/64-inch drill bit. The 30 to 35 psi in a tank sprayer will move liquid insecticide in sound wood. Using 5 to 10 1-second applications (depressions of the valve trigger) will deliver a sufficient amount of insecticide into the wood for movement along the tracheids.

    At the perimeter of the infested wood, space the injection holes about 8 inches apart on the wide face of the wood. This spacing will provide for sufficient overlap of the insecticide from adjacent treatment holes. Drill and inject two or threes sites across the width of the lumber.

Surface Treatment: Wood-Boring Beetles. Liquid insecticide applied to the surface of structural wood penetrates only the outer layer, about 1/8 inch. Surface residue from treatments dissipates within a few months because of chemical degradation and volatile loss. The greatest volatile loss occurs in house attics due to seasonally high temperatures.

Although surface treatment does not penetrate to provide immediate control of insects deep in the wood, it can provide long-term protection against future infestation. The residue immediately below the surface can control adult insects as they emerge from

Old house borer

the wood. This residue can protect against re-infestations by killing first-stage larvae that enter the wood after emerging from the egg.

When larvae are full grown, they tunnel to the wood surface to create a pupal chamber. The process of tunneling to the surface and enlarging the tunnel into a pupal chamber brings larvae into contact with the lethal insecticide residue established below the treated surface.

- *Application.* Product labels on insecticides for wood-infesting insect control recommend applying to the point prior to run-off. This is generally considered to be before excess liquid drips from the surface. The objective is to retain a large amount of liquid on the surface to ensure penetration below the surface. There is little or no advantage to spraying to excess and certainly not to the point of run-off, at which insecticide is dripping from the surface.

Approximately 1/8 inch below the treated surface, there is no significant difference in the amount of insecticide whether application is made to the point of run-off, at 100 times less, or at 50 times less than run-off.

Solvent- or oil-based sprays and formulations are sometimes used to treat structural wood. The apparent absorbency of oil into bare wood is the basis of this practice. However, there is no significant difference between penetrations of water- or oil-based sprays in structural wood.

- *Borates.* The effectiveness of surface applications of borates depends on penetration of a lethal amount of boron into the wood. In pine, residues of 1,000 to 1,900 ppm boron are lethal to first-stage larvae of the old house borer and anobiid beetle that feed on the wood. Following surface application at product label rates, lethal concentrations of 2,530 to 3,052 ppm of boron are found at the 1/32-inch depth. This amount of boron is lethal to beetle larvae that eat this treated wood.

Beetle larvae must ingest lethal quantities of boron, not simply contact the treated wood.

Wood Moisture, Rough Surface, and Flooding

- *Moisture content* of structural lumber ranges from 15% in late summer to 7% in winter. Professional treatment of wood in attics and crawlspaces occurs at any time of the year, with wood moisture at or between these extremes. Fortunately, wood-moisture content does not influence the penetration of liquid insecticide applied to the surface.

- *Rough (surface)* structural wood has greater surface area than lumber that is smooth. Rough-surface wood has a higher degree of wettability, and absorbs significantly more liquid at the point of application. This increased surface loading can result in greater penetration below the surface.

- *Rainfall.* Borate-treated wood may be exposed to normal rainfall or unanticipated flooding, both of which can threaten the stability of the residue because of the water-solubility of boron. Exposure to 0.5 inch of water (e.g., rainfall) per day for 28 days can significantly decrease boron residue at the wood surface. However, the residue remaining 1/32 inch and 1/16 inch below is lethal to powderpost-beetle larvae. While some of the surface residue may be carried away by rainfall, some of it is carried into the wood to form a lethal residue.

- *Flooding.* Treated wood may be immersed in water for short or long periods during a flood. This type of continuous exposure is a threat to borate residue on the surface and below. Immersion of borate-treated wood for 72 hours, 60 days after treatment, can result in a significant decrease in the boron at the surface and at depths below the surface. The concentration of the remaining residue is not sufficient for protection against wood-infesting beetles.

Label directions and application volume recommendations for wood-treatment insecticides may not consider the surface condition of the wood. Rough-surface wood may require a larger volume of liquid to achieve complete coverage.

CHAPTER 13

STRUCTURAL FUMIGATION

Fumigation is not a commonly used control strategy for household and structural pests, although it was once routinely used for wood-infesting beetle and drywood termite infestations. Today, chemical control methods, such as surface application and wood injection of liquid insecticides, provide effective elimination of infestations when they are limited or when the infested timbers can be identified. Occasionally, infestations are extensive or inaccessible, and control with a fumigant gas is warranted. These criteria can also apply to household pests, such as German cockroaches and bed bugs. Fumigation also remains viable for control of stored-grain, stored-food, and tobacco pests.

MODERN FUMIGANTS

The chemicals commonly used for household and structural fumigation are methyl bromide, phosphine (hydrogen phosphide), and sulfuryl fluoride. Modern fumigants have unique means of toxicity, and chemical and physical properties that make them suitable for a specific pest, structural substrate, or environmental condition. Detailed coverage of the fumigants previously and currently used worldwide, and their application methods can be found in the Manual of Fumigation for Insect Control, which is available at www.fao.org (search for Manual of Fumigation).

Environmental Effects

- *Methyl bromide.* Methyl bromide gas has been identified as one of the chemicals that contributes to degradation of the ozone layer in the earth's atmosphere. The ozone layer is critical for protecting living organisms on the earth's surface from harmful ultraviolet radiation. Parties to the 1992 Montreal Protocol on Substances that Deplete the Ozone Layer considered methyl bromide to be an ozone-depleting substance that requires regulation and an eventual worldwide ban. However, the protocol allowed an exemption for quarantine and pre-shipment applications.

- *Sulfuryl fluoride.* When a treated structure is aerated, the fumigant sulfuryl fluoride is released and rapidly dissipates into the atmosphere. Ultraviolet radiation and reactions with solid particles in the atmosphere catalyze the breakdown of this gas. Sulfuryl fluoride breaks down to release fluoride and fluorosulphate ions. The relatively small amounts of gas released are expected to have virtually no effect on the atmosphere or the global environment, and are not expected to contribute to acid rain. However, the breakdown product, fluoride, can accumulate in some food materials treated with this fumigant, and the Environmental Protection Agency (EPA) may limit the use of sulfuryl fluoride in foods.

- *Phosphine.* This gas is a natural constituent of the atmosphere at very low and variable concentrations. Atmospheric phosphine originates primarily from bacterial reduction of phosphate in decaying organic matter. The release of the relatively small amounts of phosphine following structural or agricultural fumigation is not expected to significantly change its concentration in the immediate environment. Atmospheric phosphine reacts with water and is broken down into other products.

For the egg stage of insects, the fumigant must pass through the shell, which makes this stage difficult to control. Killing the eggs of some pests requires very high dosages, which may not be economically practical.

What is released into the soil is broken down very quickly, leaving no residue. This chemical does not accumulate in the food chain.

Toxicity. The lethal effect of fumigants is primarily caused by the gas entering the respiratory system. For vertebrates, the route of exposure is through the lungs during breathing. For insects, the route of exposure is through the series of tubes (trachea) that open to the outside and carry air throughout the body. Higher dosages required to control the egg stage can be obtained by increasing the exposure time, increasing the concentration, or both.

The mode of action of sulfuryl fluoride may delay death of insects for several days following fumigation. Insects exposed to a lethal dose may be active immediately following fumigation.

- *Methyl bromide.* This fumigant is highly soluble in lipids, a major component of cell membranes such as those around nerves. Methyl bromide controls a variety of plants and animals. This may be linked to its ability to incapacitate the enzymes involved in regulating cellular metabolism. It also adversely affects the central nervous system of animals, but this mechanism does not explain its toxicity to bacteria, fungi, and seeds.

- *Sulfuryl fluoride.* Pupae are more tolerant than other life stages to methyl bromide because it binds to the pupal case, limiting the amount of chemical that contacts the pupa. However, sulfuryl fluoride prevents insects from metabolizing the stored fats they need to maintain a source of energy for survival. It has been demonstrated to reduce oxygen uptake in insect eggs.

- *Phosphine.* This common fumigant is highly toxic as it affects the oxidative respiration of animals, but it is nontoxic to organisms that respire anerobically (such as fermenting bacteria). Phosphine is reactive with oxygen, and its toxicity results from its oxidation in insect cells. For this reason, oxygen must be present in adequate amounts for phosphine to be toxic. In sites with reduced oxygen, there may be low mortality from phosphine. Any resistance is due to reduced uptake or exclusion of phosphine through the cell membrane.

Wood covered with latex paint, plastic, or other materials is not easily penetrated.

Diffusion and Penetration. Movement of gas molecules is governed by characteristics of each gas and forces of equilibrium. The speed of movement of gas molecules depends on their molecular weight. Fumigants with high molecular weight, such as sulfuryl fluoride, move slower than those with lower molecular weight, such as phosphine. During fumigation, gases move, or "diffuse," from regions of high concentration to regions of lesser concentration until there is equilibrium. A fumigant gas will spread from the point(s) of introduction to the surrounding space or rooms. It will also diffuse to the outside air unless confined by tarps or plastic sheets.

- *Temperature.* The rate of fumigant-gas diffusion and penetration of substrates is fastest at high temperatures. Cold air is heavier than warm air, and the introduction of liquid fumigant can produce a cool layer of fumigant-laden air sinking below a layer of warm air, resulting in uneven concentrations in the surrounding space. Fans must be used to mix some fumigants to prevent layering of the gas in the treated space. Once the fumigant gas establishes equilibrium in the space, it will not separate or stratify.

- *Substrates.* Fumigant gas molecules bind to the surface and penetrate nearly all materials in typical buildings. The gas penetrates the microscopic air spaces of materials such as fabric and wood; the natural pores in structural wood facilitate gas penetration. Galleries of termites, wood-infesting beetles, and carpenter ants are usually close to the wood surface and easily penetrated.

FUMIGATION SITE CONDITIONS: HEAT, COLD, WIND

Fumigants behave differently under different conditions, but ambient temperature has the greatest influence. Temperature affects both the physiological activity of the target pest(s), and the distribution and penetration of the fumigant gas. Product labels consider the effects of temperature and recommend different dosages for different temperatures. In warm conditions, the fumigant gas molecules move rapidly and diffuse throughout the treated space. Labels for fumigants generally require a minimum air temperature at the time of fumigation.

Below about 50°F, most insects have limited respiration, and their metabolic activity is slowed enough that they may not be killed by a typically lethal concentration of fumigant gas.

Temperature. Cold conditions affect fumigant gas molecules by decreasing their movement, and, as a result, they may not diffuse evenly through the treatment space. If this results in the survival of even just a few individuals, the pest infestation may return to previous levels. Insects and other arthropods are cold blooded, and their body temperature is linked to the surrounding conditions. Respiration increases with an increase in temperature. Increased respiration usually results in greater penetration of the insect cuticle and uptake of the fumigant gas.

Wind. The size of the building, as it relates to surface area to volume, can influence the loss of gas due to wind, especially from tarped buildings. As the size of the structure increases, the volume grows much faster than the surface area of the tarps covering it. The ratio of tarp to surface area compared to volume is high for small buildings but less for large buildings.

In large buildings, there is a large volume of gas but a relatively small surface area for the gas to escape. In small buildings, there is a small volume of gas but a large surface area with greater potential for escape.

FUMIGATION DOSAGE AND CONCENTRATION

The goal of fumigation is to achieve and maintain a lethal dosage of gas in the treatment space. The lethal dosage is accomplished by confining a sufficient quantity or concentration of gas for a specific time to kill the target pest.

The effective, lethal fumigation dosages are calculated as the product of concentration (C) x exposure time (T): Dosage = C x T. Exposure time is usually measured in hours and concentration in ounces per 1,000 cubic feet. In general, there is a direct relationship between dosage and the resulting toxicity in insects. The effective dosage for a given insect pest can be achieved by varying either the concentration or time.

FUMIGANT GAS CONFINEMENT

Successful fumigation depends on how well the space or structure is sealed. The toxic gas must be confined long enough for the pest to receive a lethal exposure. The better the seal, the less gas and/or time needed to achieve the required dosage. Fumigants quickly move through plasterboard and wood walls; even concrete block and brick walls permit loss of fumigant gases. Almost any structure can be modified to hold gases, but some residential buildings must be covered with polyethylene or vinyl-coated nylon tarpaulins to ensure control.

Half-Loss Time. The rate at which the gas escapes the treatment space is referred to as the "half-loss time." It is the time required for the loss of half the gas concentration during the fumigation time. Most fumigated structures leak gas to some extent. This can be influenced by wind, building construction, and the degree of sealing. Gas concentration can also be affected by sorption into contents of the treatment space. As fumigant escapes the structure, there is a continuously decreasing amount of gas confined against the target pest. To compensate for the fumigant lost, a greater initial concentration, longer period of exposure, and reintroduction of fumigant is used to attain the lethal dosage.

Tarp Covering. Tarpaulins can reduce the above-ground loss, but a significant amount of gas can escape through the ground seal and/or the under-seal of the building. Procedures to improve the ground seal include spraying the soil with water and using sand- or gravel-filled flexible tubes to secure the bottom of tarpaulins. Tarpaulins or polyethylene sheets can be used to cover the soil in accessible areas to reduce gas loss when fumigating buildings with crawlspaces.

Tape and Seal uses polyethylene sheets and tape to prevent leaking around doors, windows, vents, and other openings. Caulking and other means are used to seal small openings, cracks, and joints. Plastic tape and sealing compounds can be used to confine fumigant in small buildings, trucks, shipping containers, and structures constructed of metal, brick, or concrete. The tape-and-seal method is less effective than tarping.

Container. Small cargo or commercial moving trucks may serve as fumigation chambers for household furniture and other infested materials. Large-volume vehicles with detached cabs can be effectively sealed with plastic tape, applying 4-mil polyethylene sheets to seal the rear rolling door, and taping the bottom of the frame around the rear door. It is necessary to monitor the fumigation to determine the half-loss time and dosage accumulation.

Fumigant gas permissible exposure levels (PEL) are 5 ppm for sulfuryl fluoride and methyl bromide; 3 ppm for phosphine; and 0.1 ppm for chloropicrin. The PEL is considered to be substantially lower than the concentration that may affect people following long-term exposure.

AERATION AND RE-ENTRY

Aeration is the final step in structural fumigation. It involves allowing the fumigant to dissipate in the outside air, where it is diluted to a harmless concentration. This process involves removing the tarps, plastic sheets, or sealing tape from the building, and opening doors and windows. Fans may be used to ventilate and clear the building of the remaining gas.

Re-entry. Once the building has been aerated, the level of remaining fumigant gas is measured to ensure that it is below the approved concentration for re-entry. Specifically designed equipment is used to determine gas concentrations inside buildings. Typically, the detectable levels of gas are less than1 ppm within 6 hours of the start of aeration, and there are no detectable levels 24 hours after the start of aeration.

Extremely low levels of fumigant can remain for a short time in air spaces between walls and inside cabinets and in upholstered furniture. Gas in these areas will continue to dissipate for a few hours.

REGULATIONS AND SAFETY

All structural fumigants allowed for use in the U.S. must be registered by the EPA, and may be further regulated under state laws. These regulations include the use of chloropicrin, clearing devices, and self-contained breathing apparatuses, and the placement of warning signs. There are significant differences in the product-label instructions and restrictions among fumigants.

Fumigants are classified as restricted-use pesticides, for use only by certified applicators and persons trained in the proper use of fumigant gases. Manufacturers may also require special training beyond government requirements for the purchase of a fumigant.

Exposure Levels. Lethal poisoning of insects is based on exposure time and accumulated dosage. There are no direct experiments to determine the human response to fumigant gases. However, human exposure to low concentrations for long periods and high concentrations for short periods can be toxic.

Respiration Protection. Inhalation of a fumigant gas is the primary route of exposure for humans. Respirators or self-contained breathing units, are required protection for operators.

Workers are required to wear positive-pressure self-contained breathing apparatus (SCBA) gear when entering areas where concentrations of gas exceed the permissible exposure level or areas where concentrations are unknown.

SCBA respiratory gear is designed for exposure to toxic gases; ***SCUBA*** (self-contained ***underwater*** breathing apparatus) systems are designed for underwater use, not for exposure to fumigant gases.

STRUCTURAL FUMIGANTS

The differences between modern fumigants are based on: the temperature at which the fumigant changes from liquid to gas; the dispersal of the fumigant throughout the structure; and the weight of the gas in relation to air, with some being heavier and some lighter than air. They are all effective insecticides, but vary in their control performance.

Methyl Bromide. This fumigant is generally limited to quarantine and pre-shipment treatments and granted critical-use exemptions. Methyl bromide is formulated as a liquid with or without the warning agent chloropicrin. It has a specific gravity 3.3 times heavier than air, and fans are needed for distribution and equilibrium of the gas in the structure. Methyl bromide can kill all insect life stages in 2 to 24 hours.

Methyl bromide will break down to a corrosive compound in the presence of an open flame, pilot light, or other intense heat source. It reacts chemically with certain materials, especially sulfur-containing products, such as foam rubber and furs.

Phosphine. Phosphine gas, or hydrogen phosphide, is used primarily for fumigation of grain bins, seed, bulk stored flour, dried fruits, nuts, and other food products. It is not as fast-acting as is methyl bromide or sulfuryl fluoride. Active stages, such as adults and larvae, are killed rapidly, but insect eggs and pupae that have lower respiration rates can survive long periods with minimal respiration.

Phosphine can react with copper, gold, and silver to form an oxide residue. It can damage electrical wiring, electrical appliances, telephone equipment, computers, and microprocessor-controlled machinery.

Sulfuryl Fluoride. Sulfuryl fluoride is used primarily in structural fumigation, but recently has found use in container fumigation for home furnishings infested with bed bugs. It is formulated as a liquid under pressure in steel cylinders.

Since sulfuryl fluoride is odorless at working concentrations, chloropicrin gas is applied as a warning agent before structural fumigations. Sulfuryl fluoride is not formulated with chloropicrin because the two chemicals do not release from the cylinder homogeneously. Sulfuryl fluoride is heavier than air, and fans are required during introduction to mix air chilled by the volatilization of this gas.

Sulfuryl fluoride is nonflammable, but in the presence of a pilot light, it will decompose to form hydrofluoric acid. This acid is very corrosive and may etch glass or tile or tarnish and pit metal.

Chloropicrin. Chloropicrin is most commonly added as a warning agent to odorless gases used in structural fumigation. It is often formulated with methyl bromide in cylinders, and it is packaged separately as a liquid to be used during sulfuryl-fluoride and methyl-bromide fumigations.

Chloropicrin causes eye irritation and tears at very low concentrations of 0.3 ppm; eyes will close at 3.0 ppm; and an individual will become incapacitated at 4.0 ppm. Chloropicrin absorbs readily into organic matter and desorbs slowly.

CHAPTER 14

INDOOR INSECTICIDE APPLICATION

Indoor application of insecticides requires knowledge of the target pest and the insecticide, accurate and reliable application equipment, and consideration of the safety of people and pets. The objective of most applications is to create an insecticide residue on a substrate that will be contacted by the target pest. Success is linked to the interaction of the pest and insecticide within a limited time. This will depend on characteristics of the active ingredient, the formulation, and the substrate.

APPLICATION METHODS

Liquid Spray. The majority of insecticides for household pests are applied as a liquid spray. Fine or coarse fan sprays and crack-and-crevice sprays are used to treat substrates and concealed harborages. The terms "fine" and "coarse," as applied to the spray, refer to the amount of liquid delivered and the size of the droplets: coarse fan sprays deliver more liquid with larger drops than do fine fan sprays.

- *Spray pattern and overlapping.* The distribution of liquid in a fan-spray pattern is concentrated in the center of the swath, with less at the edges. Old and worn nozzles may not create this distribution, instead the pattern may be distorted. Standard distribution becomes important when surfaces must be treated with insecticide to create a uniform residue. It is important, for example, when treating the foundation wall and adjacent soil of a building during perimeter application or when treating carpeting for fleas.

  To achieve uniform coverage of the surface, the spray pattern must be overlapped or extended into the area of the adjacent pattern by about 7 inches. This will help eliminate the decreased amount at the edge of each swath. Applying an edge-to-edge spray pattern onto surfaces can result in untreated areas or areas with sub-lethal residue where the pattern edges meet.

Nozzle distribution pattern

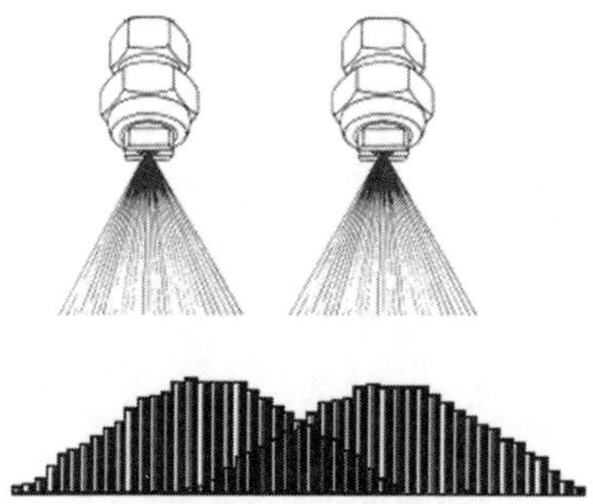

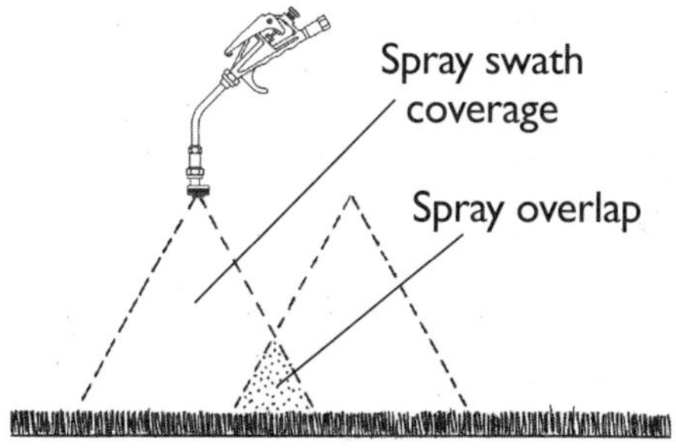

- *Spot treatment.* A spot treatment is defined as the spraying of a residual insecticide to a limited-sized area. It differs from a crack-and-crevice application because application is to an exposed surface rather than into a crack. It differs from a general application because the size of the spot cannot exceed 2 square feet. However, the dimensions of the actual spot can vary: it could be a 1- by 2-foot rectangle or a 6-inch by 4-foot strip, or it may be a circle with a 19-inch diameter.

Spot application is often prescribed in areas where food is prepared, or the product label does not permit general spraying.

Product labels may prescribe the size and number of treated spots, especially in food-handling facilities. For example, if spots are 10 square feet, the limit may be 50 in 1,000 square feet of floor space.

- *Crack and crevice.* Any opening which may lead to void areas is considered to be a crack or crevice. As an application method, crack-and-crevice treatment is used in control programs for crawling pests, but it is most effective for cockroaches.

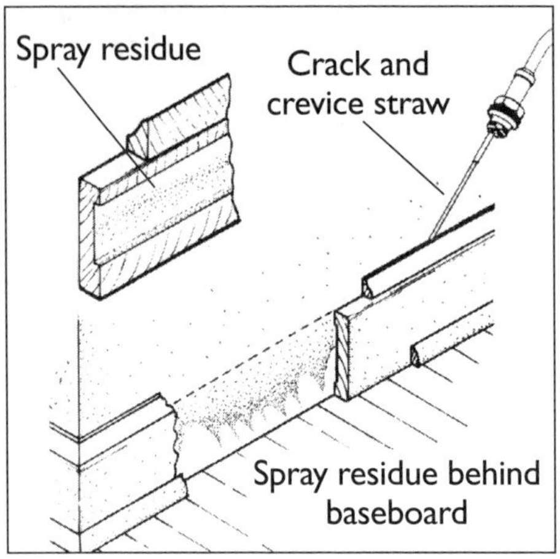

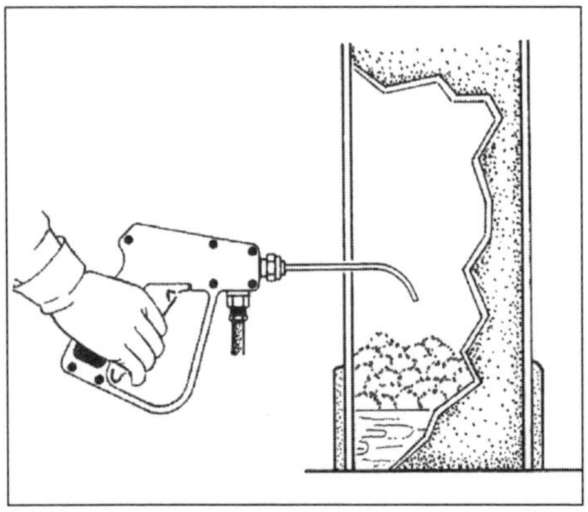

Foam application to wall void

Crack-and-crevice treatments are usually made with a tip extension on the nozzle. The most common is a 6-inch, flexible-plastic tip. These tips, or straws, are more effective than fan sprays in treating cracks infested with German cockroaches. Control of this pest increases when crack-and-crevice straws are used and inserted at least 1/4 inch into a crack during application. Moving the straw along the length of a crack while spraying improves the coverage of liquid insecticide on surfaces contacted by cockroaches and other pests.

- *Injection.* Spraying or foaming into structural voids and insect galleries is effective in controlling wood-infesting insects. A wood-injection nozzle can be used to inject insecticide into the gallery through a hole drilled in the surface. Voids between wall studs are treated similarly, with the injection hole drilled near the baseboard, and the liquid or foam directed at the sill plate and adjacent framing-stud bases.

Foam. The advantages of delivering a liquid insecticide as foam are the abilities to treat concealed or difficult-to-reach spaces and large voids and to establish a continuous and uniform insecticide residue. Because the nature of foam is to move across uneven surfaces and surround objects in its path, it is more effective than a liquid spray in some locations. In addition to application in wall voids and spaces beneath cabinets, foam is commonly used to treat the space under concrete slabs with termiticide.

Fogging and Aerosols. Professional pest-control use of aerosols began as crack-and-crevice treatments, but now includes treating of exposed surfaces for crawling pests. Professional aerosol-delivery systems enable precision application by having the actuator extended from the can by a coiled hose. This enables the user to reach and treat crevices that would be otherwise inaccessible with the aerosol can alone.

- *Droplets.* Oil- and water-based formulations can produce different aerial concentrations of droplets. Fogging with water-based insecticides is more effective than with oil based. The large droplets of oil-based material fall to the floor immediately after being sprayed, leaving a limited number of small droplets in the air. However, with water-based formulations, the large droplets shrink from evaporation. As a result, the shrunken droplets remain in the air along with the original small droplets, providing a higher aerial concentration of insecticide.

- *Cold fogging.* Cold-fogging machines produce droplets larger than 50 microns in diameter. These foggers are commonly used to treat enclosed spaces for the purpose of flushing or killing insect pests. The most effective droplet size for killing house flies is 30 to 35 microns, and about 16 microns for fruit flies. Cold foggers are sometimes used as a space spray following the application of liquid sprays for control of crawling insects, such as cockroaches. However, most droplets contact exposed surfaces, with few carried into narrow voids or harborages.

Misuse of total-release aerosols can result in explosion, fire, and serious health problems. Consumers may use an excessive number of these foggers, then reoccupy the treated area before proper ventilation.

- *Total-release foggers.* Total-release aerosol cans are considered to be foggers and are used for treating large indoor spaces. However, the effectiveness of this application method is limited. The droplets produced are large and fall quickly, within 30 seconds. Few if any droplets are carried to crevices to kill cockroaches or other crawling insects, or to the base of carpeting to kill flea larvae. These foggers are designed to treat unobstructed space. The treatment space is recommended on the label; treatment of smaller spaces is not permitted.

Baits and Lures. Toxic baits in the form of liquid, gel, dust, or granules are a control strategy for household pests. Other forms of toxic and non-toxic bait are available for termites. The efficacy of baits for cockroaches and ants is based on application, so that a large number of individuals find and eat the bait. Killing a large percentage of cockroaches in a population, especially females with egg cases, can result in decline and elimination of the infestation. When a large number of foraging ants return to the nest and share toxic food with nestmates and queen(s), the colony dies.

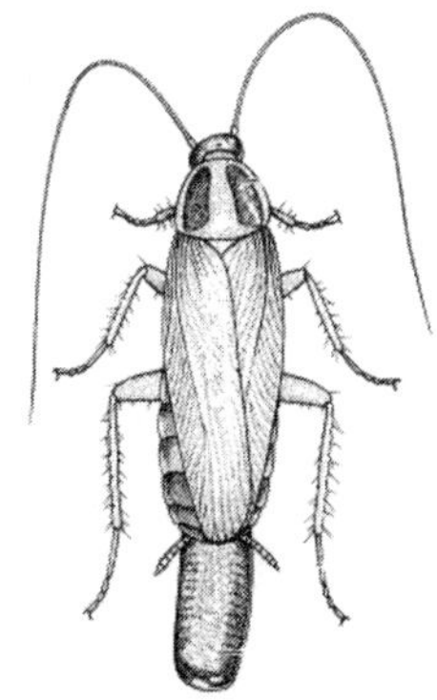

Female cockroach with egg case

- *Gel bait.* Formulations of gel bait can be considered a residual liquid applied as a crack-and-crevice or spot treatment. Moisture evaporates from gel-bait placements to leave a semi-soft, palatable residue that must be eaten; some gel-bait formulations are also toxic on contact. Gel baits are applied as a spot or extended bead in cracks and crevices, but not usually to exposed surfaces. Sites where temperatures are above 130°F are not treated because the gel ingredients may liquefy.

- *Lures.* Food lures and attractants are used in sticky traps for cockroaches and house flies and in capture traps for fruit flies, house flies, and wasps. Traps for blood-feeding insects, such as bed bugs and mosquitoes, include a combination of carbon dioxide and volatile chemicals that are attractive to adults searching for a blood meal. Baited traps are practical monitors of pest infestations and are useful in detection, determination of distribution, and documentation of pest presence.

Dust and Granules. Dusts are usually applied to pest harborages, wall voids, baseboards, nests, and other sites inaccessible from the outside. Application is made with bulb dusters, power dusters, and small- or large-capacity hand dusters. Large power dusters may be used for treating wasp and bee nests, as well as large spaces, such as attics and crawlspaces. Proper application avoids exposed surface deposits and airborne dust. The direct transfer of dust to the pest makes it effective against a wide range of crawling insects. For wood-infesting insects, dust is applied to galleries for immediate and long-term control.

Granular formulations for indoor pest control usually have label limitations for use only in areas that are inaccessible to children and pets. Typically treated sites include behind electrical switch plates and in wall voids, attics, and crawlspaces. Application is best made with a hand or bulb duster.

ENVIRONMENTAL CONDITIONS

Indoor conditions are extremely variable and make this one of the most difficult locations to apply insecticides. Household surface temperatures have a minor influence on insecticide residues, since most formulations are stable at the temperatures encountered indoors. However, many insects are influenced by low relative humidity; for example, cat flea larvae do not survive if the humidity is 33% or lower.

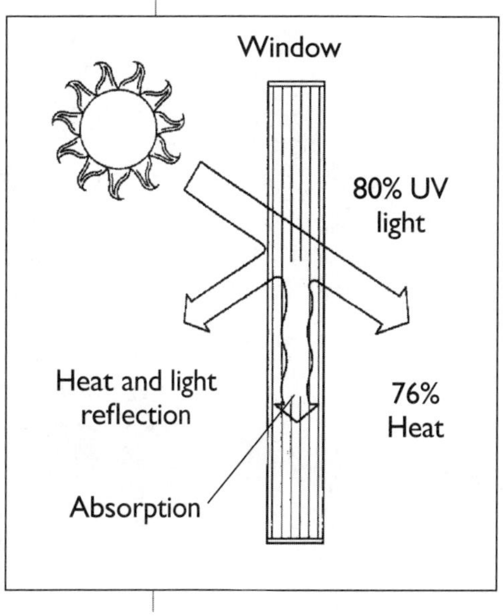

Ultraviolet Light. Window glass removes a small amount of the ultraviolet (UV) portion of the light spectrum, but the band between 325 nm and 400 nm gets through. It is this portion of visible light that degrades insecticides and can affect the performance of residues. This is also the portion of UV light that is attractive to insects, especially flies, which is why they fly to windows.

The UV portion of sunlight striking treated carpet, ceramic tile, or other surfaces may reduce the activity of insecticide residues. However, the stability of some modern insecticides is extended, because they have ingredients that protect them from UV light.

Moisture and Decay Fungi. Water leaks in wall voids and enclosed spaces in kitchens and bathrooms can lead to damage by fungi. These spaces are often inaccessible, and detection of a problem before it results in structural damage is difficult. Plywood, particle board, and chipboard are used as under-layment for flooring and cabinet framing. Much of this wood is susceptible to fungal decay if exposed to prolonged moisture. Wood swelling caused by moisture uptake, coupled with glue-bond failures can cause a loss of strength.

Moisture-damaged wood is attractive to termites, carpenter ants, and other insects that use it as food or harborage. Some decay fungi produce volatile substances that can attract termites.

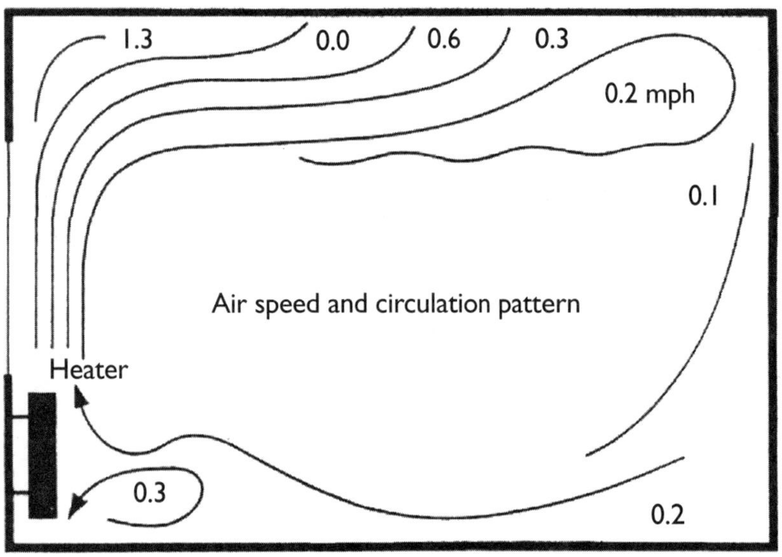

Air Currents. There is a normal amount of air circulation in rooms, typically generated by the movement of air from a wall-mounted heating or cooling unit or an open window. Although the air speeds are low, there is usually a consistent movement of air from floor to ceiling. This air circulation can influence the distribution of small droplets created from fine and coarse fan sprays.

The forced air from ventilation, air-conditioning, and heating systems is distributed around rooms from ceiling, wall, or floor-mounted vents. Air discharged from a floor- or wall-mounted grill with vertical vanes will spread air horizontally and vertically.

Ceiling vents with radial and rectangular diffusers have a shorter throw or spread distance than slot diffusers. These are common in commercial buildings. They provide a high degree of air diffusion and mixing. Airflow from these vents is typically directed toward the walls. In a room with a 10-foot ceiling, the airflow from a ceiling vent will contact the wall at about 6 feet high, and from there, the warm or cold air circulates below.

Before space spraying or cold fogging indoors, the air currents from floor and ceiling vents must be considered. It is best if these systems are not active at the time.

SURFACES AND SUBSTRATES

The majority of indoor insecticide application is made to porous surfaces, such as unfinished wood, latex-painted wood, natural and synthetic fabrics, and carpeting. The non-porous surfaces treated indoors are few, and they may or may not be conducive to insecticide residue.

Porous Surfaces

- *Fabric.* Synthetic fabrics used on furniture are suitable for retaining an insecticide residue at the surface. The most effective residue is achieved through a spray of fine droplets that does not wet the surface. Cotton twill is commonly used as a furniture covering. It can retain insecticide solution within the fabric spaces, rather than on the surface.

- *Carpet.* Carpet composition and pile height influence penetration of insecticides applied to the surface. A fine fan spray to carpet results in penetration to only about 1/3 of the pile length, with 93% of the insecticide residue remaining in the upper third of the carpet pile. Increasing the application pressure or decreasing spray height does not increase penetration.

Furniture fabric treated with stain or water-repellent materials may permit little or no insecticide penetration.

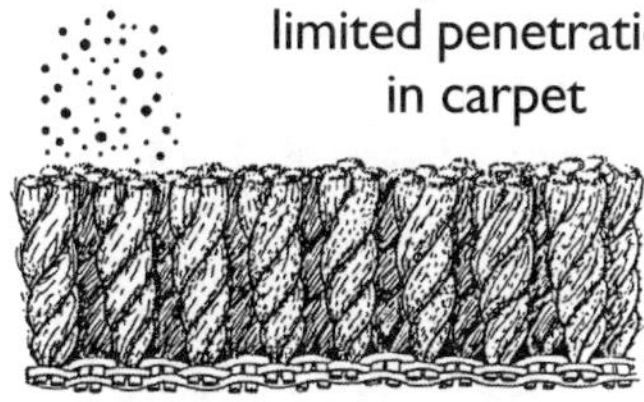

Contrary to common belief, high application pressure does not deliver more liquid to the carpet.

The most common carpet treatment is directed at the larvae of the cat flea. Flea larvae spend 80% of their time at the base of the carpet pile, while adult fleas spend the majority of their time on the animal. Flea control requires treatment of the host animal to kill adult fleas, and application of residual insecticides to the carpet to control larvae.

- *Wood.* Indoor exposed-wood surfaces are usually finished with paint, varnish, or stain. There are few indoor locations with bare or unfinished wood, these include the interior of cabinets, drawers, and voids behind and beneath kitchen and bathroom counters. When treated with liquid insecticide or foam, these surfaces are capable of maintaining a surface residue and absorption to slightly below the surface. Finished surfaces resist liquid penetration, although if the actual surface is a matrix, it may be somewhat porous.

Non-Porous Surfaces

- *Stainless steel.* Stainless-steel surfaces are common in kitchens and restaurants. Equipment treated with insecticide may include areas of stainless-steel tables, storage bins, sinks, and cabinets. Almost all metal surfaces, including stainless steel, are covered with an oxide layer. The chromium in stainless steel forms an invisible layer of chromium oxide on the surface.

The corrosive oxide layer on the surface of stainless steel can significantly reduce the adhesion and stability of liquid insecticides.

- *Ceramic tile.* Ceramic tile in residential and commercial buildings is found primarily in kitchens and bathrooms. The surface of tile can be difficult to treat with liquid insecticide, as it provides little adhesion for droplets and over application may result in run-off. Residual activity of insecticides on tile is long if undisturbed. In commercial kitchens, food-manufacturing plants, and health-care facilities, tile flooring is frequently cleaned with detergents and disinfectants, which removes insecticide residue.

SENSITIVE AREAS

Locations that are considered to be sensitive include hospitals; veterinary, medical, and health-care facilities; schools; food-processing plants; pet stores; and food warehouses. Sensitive sites within buildings include rooms for medical supplies, electronics, and computers; animal cages; fish tanks; and pet-food storage rooms.

These sensitive treatment areas require careful use of conventional and alternative control methods.

Baits. Insecticides formulated as baits provide limited insecticidal exposure to people or pets. Gel baits can be effective when used with a sticky-trap program to locate cockroach-infested harborages. The use of only a few bait placements may be enough to eliminate a small cockroach problem around electrical or medical equipment.

Professional sticky trap

Traps. When placed close to infested harborages, sticky traps can be an effective control tool. These traps can reduce infestations by removing males, females, and nymphs. Traps may not always provide long-term control, but they could help limit the need for insecticide application to a few locations.

Where rodenticides are prohibited, snap traps, sticky traps, and glue boards/trays can provide control. These capture traps should be used in a covered station, which protects the trap bait and glue surface from tampering and debris and shields the captured rodent from public view.

SENSITIVE PEOPLE

Individuals with ongoing respiratory problems, various allergies, or a history of sensitivity to airborne chemicals may also be sensitive to insecticides applied in their immediate surroundings. Surface spraying and some crack-and-crevice treatments can trigger allergic reaction. Reduced application and alternative methods may be necessary to accommodate these individuals, while still providing control of the infestation. Notification of the treatment schedule and type of insecticides used may help reduce apprehension.

Asthma is a respiratory reaction to specific material(s) or allergen(s) in the environment. Environmental allergens include airborne mold and substances such as insecticides and chemical solvents.

Allergies and Asthma. Individuals in offices, apartment buildings, and residences may object to insecticide application on the basis of allergies to insecticides or other chemicals. An allergy is a hypersensitive reaction to exposure to a chemical or compound foreign to the body; about 20% of the U.S. population have allergy or asthma symptoms, and 55% test positive to one or more allergies.

The most common allergies are to pollen, pets, and food products or ingredients. However, the term "allergy" has become popular and commonly used in modern society. Many people who are sensitive to something think they are allergic to it, but the basis of the allergy has not been medically established.

Insecticides usually consist of two components: the active ingredients, which make up a low percentage of the overall product, and the inert ingredients, which account for the bulk of the product. The active ingredient is a highly purified component and, in most cases, has no odor. The inert ingredients consist of petroleum distillates and emulsifiers; some people react to the strong odor of petroleum distillates. A person with a severe respiratory response to insecticide would probably be hypersensitive to one of the inert ingredients, rather than to the active ingredient.

Allergic symptoms last for hours or days; irritant symptoms clear up when the person moves away from the source.

Allergic Response. The reaction to bed bug, flea, and mosquito bites varies greatly among individuals. For most people the reaction to a bite is a red mark that is mildly itchy. However, some people have an allergic, hypersensitive reaction to the fluid that the insect injects into the skin prior to taking blood. Although this is normally harmless, not causing an immune system response, those who are allergic or sensitive to insect bites, can experience intense itching, swelling, redness, and prolonged irritation.

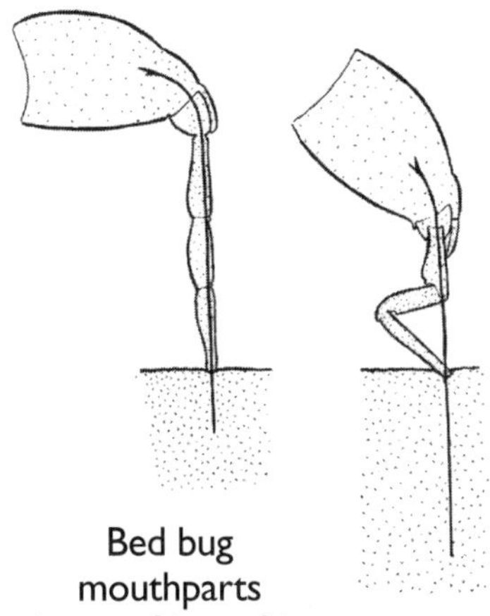

Bed bug mouthparts inserted into skin

When blood-sucking insects—such as mosquitoes, bed bugs, and cat fleas—bite, saliva with anticoagulant properties is injected into the host. This material is the basis of a person's allergic reaction to the bite.

Chemical Sensitivity. Individual's objections to insecticide application may be based on some level of chemical sensitivity. Multiple chemical sensitivity (MCS) is a condition in which symptoms occur with, or are caused by, low-level exposure to a variety of chemicals, including insecticides. The symptoms include sneezing, sore throat, breathing problems, headache, skin rash, or hives. There are several theories to explain the cause of MCS, including allergy, toxic effects, and neurological sensitization. However, there is insufficient evidence to confirm a relationship between possible causes and symptoms.

▶ Cat fleas and bed bugs have a tendency to "test bite" when on the skin of a host; this often results in a line of three or four bites in a row.

MCS has been rejected as an established organic disease by the American Academy of Allergy and Immunology and the American Medical Association.

People reporting sensitivity to insecticides will benefit from the new generation of insecticides and precision application tools. The new insecticide classes include active ingredients that are categorized as Reduced Risk and have little or no toxicity to people or pets.

HOUSEHOLD FEATURES

House Pets. Many pets have free access to the household and may be exposed to insecticides during or after application. During application, the exposure is primarily to small droplets, as droplets as large as 100 microns remain in the air for 10 seconds. This may be sufficient time to be moved by household air currents to pet cages, fish tanks, or other sites.

After application, pets may contact wet and drying residue on exposed surfaces. Prolonged contact with wet residue can result in toxic amounts of insecticide transferred to the pads of the feet and to the mouth through grooming. Product labels usually recommend that treated surfaces are dry before pets are permitted access to treated rooms.

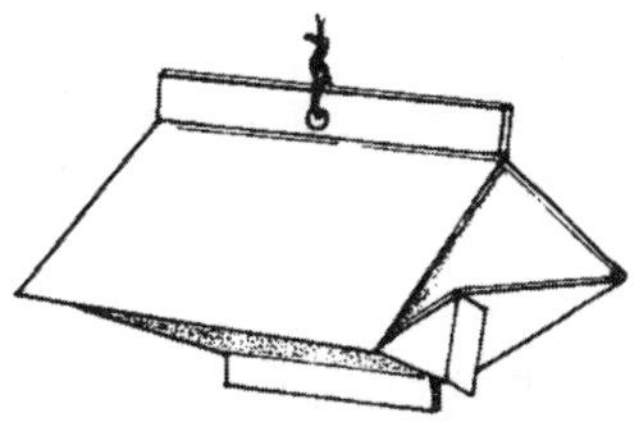

Pheromone sticky trap

Dry pet food is usually stored close to household foods. Meal moths, grain beetles, and psocids are common in pet food, and they often move into household products. Chemical control of food pests is limited by the potential of contamination. Population suppression and monitoring of grain and flour pests can be conducted with pheromone sticky traps and glue-board traps.

Furniture. When furniture is treated with liquids, dusts, and baits, fabric damage or discoloration and exposure of people and pets to insecticide residue must be considered. Natural or synthetic-fiber coverings may be susceptible to liquid insecticides, especially old furniture or those pieces exposed to sunlight.

Potted Plants. Drift or accidental exposure to insecticides may result in leaf spots or other damage to sensitive houseplants. Some formulations of household-use insecticides may be toxic to indoor plants. Small spray droplets may be carried in air streams from open windows or floor and ceiling registers to plant leaves.

Water droplets on the leaves of plants, such as African violets, may cause white or yellow patches, which are often mistaken for insecticide damage.

Books and Paper. Liquid and dust application around books and paper should be limited because of the potential of damage by spray droplets and contamination by dust. Droplets from sprays may drift to adjacent paper and cardboard packaging in warehouse, office, or residential locations. Discoloration or damage can occur to exposed paper or fabric-based materials.

Insecticide application to book shelves and storage containers may be incorrectly linked to the occurrence of brown- or rust-colored spots on old books and manuscripts. This condition, called "foxing," is caused by high humidity and temperature extremes on the ingredients of old inks.

Electric Appliances. Treatment around or into compressor motors and floor-mounted air conditioning units may include crack-and-crevice application of liquids and gel formulations of insect-control baits. Coarse fan sprays are appropriate for treating around the base of floor units. Application should be made when the circulation fans are not running to prevent small droplets from being carried in the air stream.

Moth fly

Moth fly infestations are usually caused by the accumulation of organic matter around the top of drains and further into the drain pipe.

STRUCTURAL FEATURES

Floor Drains. Insecticidal control of flies and other insects associated with floor and sink drains is limited. There are few products labeled for these sites, and the wet, organic substrate limits their efficacy. Physical repair, cleaning, and removal of organic waste are the most effective methods of controlling drain infestations of moth flies and dark-eyed fruit flies.

The bacteria in microbial drain cleaners break down organic material, eliminate volatile and attractive odors, and reduce breeding substrates.

Wall Voids. Treatment of wall voids and other large structural spaces includes the covering of exposed surfaces with insecticide to provide immediate and long-term residual control. Although there may be only limited access to the space, the application equipment must give uniform and complete coverage.

Wall voids on perimeter walls are usually filled with insulation, but voids in interior walls may be open.

- *Aerosol sprays* will give quick knockdown of pests, but the small droplets have only limited residual activity. Aerosol droplets can disperse in an empty space but are easily blocked by insulation or other material.

- *Foam* also has the ability to cover exposed surfaces and to establish a residue, with the added ability of the liquid that drains from the foam to penetrate into the wood surface and cracks and crevices. Foam is usually applied at the bottom of a wall void, immediately above the baseboard. However, for carpenter ants, the treatment point may be placed at the level of the suspected nest site. A curved tip can push back insulation in the void to allow for treatment. Moving the tip during application from one side of the void to the other will cover the exposed sill plate and the bottom of the framing studs.

- *Dust* can be applied with a hand duster, or with a pressurized duster to treat larger spaces, such as voids that do not contain insulation. Dust can cover all exposed surfaces and give both immediate and long-term protection. Applying dust to wall voids can be done at electrical- and cable-outlet boxes with a bulb duster and a plastic crack-and-crevice tip. Outlets are typically placed close to the bottom of voids, and there is access to the inside even if insulation is present.

***Caution*: To prevent electric shock, dust application to electric outlets should be done only with a rubber bulb duster fitted with a plastic tip. A metal application tip inserted into an electric outlet will conduct electricity and may cause a short circuit or electric shock.**

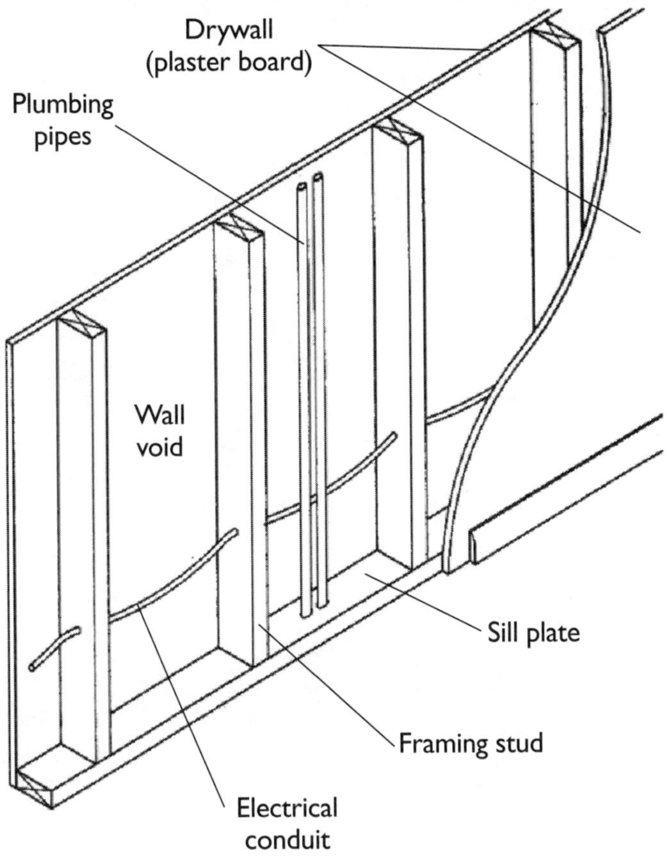

CHAPTER 15

OUTDOOR INSECTICIDE APPLICATION

Many common indoor pests originate from outdoor populations, and control applied outside may reduce or eliminate them. Successful control programs for these and other pests are based on integrating inspection and prevention strategies with insecticide application. Non-chemical methods are effective against pests that are dependent on key features of the environment. For example, subterranean termites require a source of cellulose for food; earwigs require moist and dark harborage; mosquitoes require stagnate water for breeding. Seasonal pests, however, are less likely to be controlled through habitat modification methods.

INSECTICIDE APPLICATION METHODS

Outdoor application usually involves more insecticide than does indoor application. For example, a perimeter band 5 to 10 feet wide adjacent to a structure can be treated with a coarse spray, and application volumes can be as much as 10 gallons per 1,000 square feet. An outdoor spot application, or "broadcast spot," can be as much as 4.6 pounds of granular insecticide per 1,000 square feet.

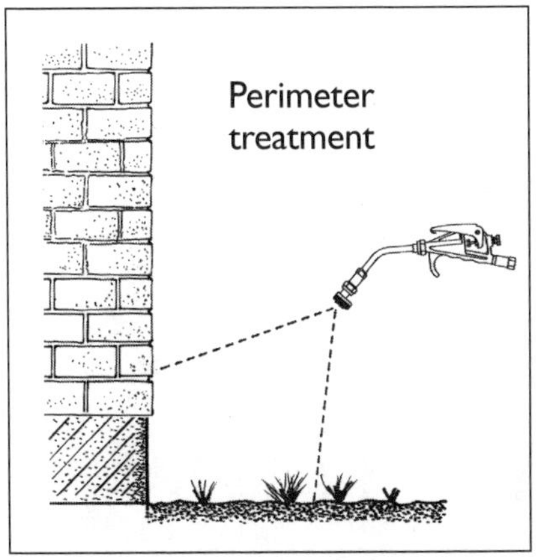

The objective of outdoor application is to treat a broad surface area to maximize the potential of contact kill and residual control to prevent pests from entering the structure.

Liquid Spray. Outdoor application is made with a tank sprayer or backpack sprayer using a coarse fan spray. This type of delivery system provides the low pressure and precise covering required by insecticide labels. The strategy is to thoroughly treat a limited area. Treating a band of soil adjacent to the foundation wall will likely contact ant nests or harborages for crickets, earwigs, or spiders. A fan-spray nozzle and application to narrow cracks and edges will be needed to treat the warm, sunny surface of the foundation wall and siding.

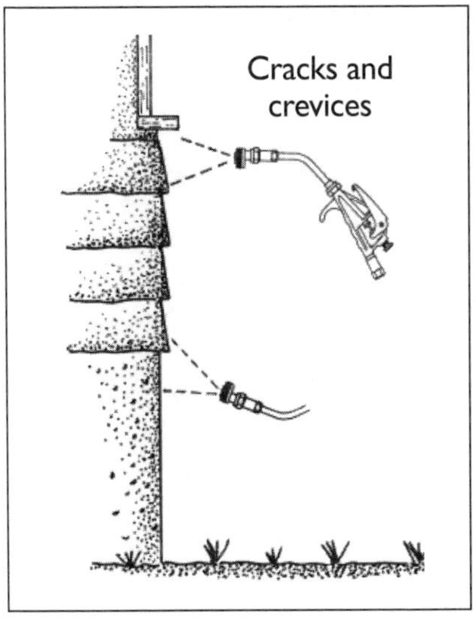

- *Cracks and crevices.* Treatment for perimeter pests involves treating of narrow cracks and crevices that act as travel routes for some pests. Ants and other insects generally travel along protected edges, such as along the top edge of the foundation or the lower edge of siding. The seams around windows, door frames, and thresholds can be treated with a fan spray or a crack-and-crevice straw. Clover mites and stink bugs move in large numbers from turfgrass to foundation walls and climb up along edges to enter window frames; these travel routes and entry points must be treated.

- *Injection.* Application of liquid into structural voids or insect galleries is an effective control method for carpenter ants and carpenter bees. Infested galleries inside trees, fencing, deck timber, and other structural members can be drilled from the surface to locate the interior gallery. A wood-injection tip can be used to inject insecticide into the gallery. Soffits and voids behind exterior siding are treated similarly, with the injection hole drilled near the nest.

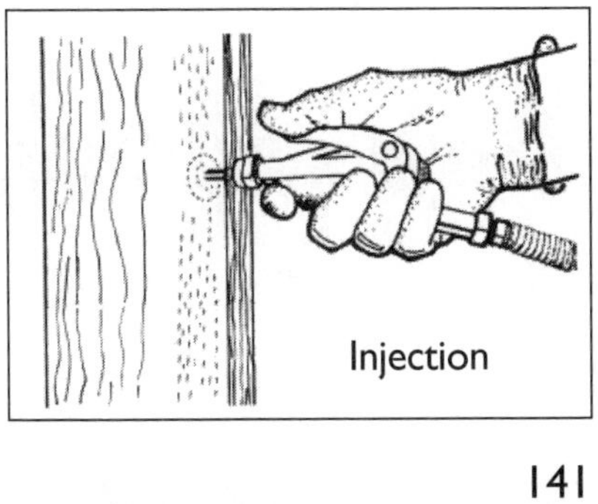

Dust. Sites for dust application include: around windows and doors, under door thresholds, at the junction of foundation wall and siding, around openings of pipes and cables that enter the foundation wall and siding, around the opening to the crawlspace, and into exposed surfaces in the crawlspace and attic. Other applications can be made into termite tubes and carpenter-ant and carpenter-bee galleries to eliminate the colonies.

Dust application in sewers provides quick knockdown of exposed cockroaches and short-term population suppression. An effective treatment can be made through the use of power equipment to disperse dust, at a rate of 1 to 2 ounces per 200 feet, through a manhole access.

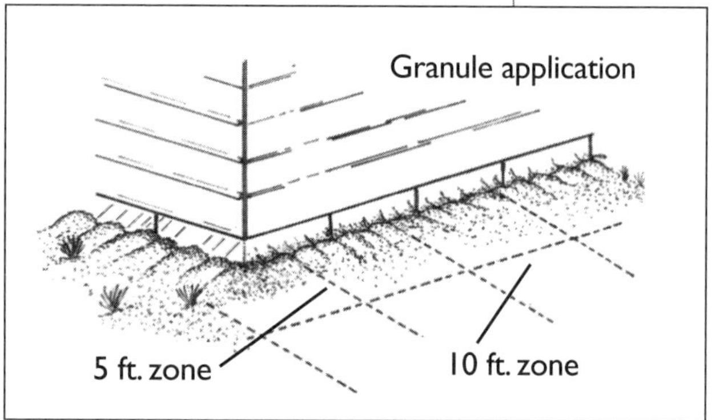

Granules. Granular formulations provide control and prevention of insects that crawl or live in the top layer of soil. Granules deliver insecticide to the soil in a small zone around and below them. Release of the active ingredient is usually triggered by exposure to moisture, thus product labels often recommend applying water after application. Application rates are generally 2 to 5 pounds per 1,000 square feet, and are designed for overlap of granule treatment zones for control of pests on the soil surface.

For subterranean termites, the purpose of granular applications is to kill termite workers foraging close to the soil surface and winged reproductives that may be on or near the soil surface.

- *Perimeter application* is defined as treating a ground area 5 to 10 feet around and adjacent to the structure. Product labels usually include guidelines based on linear distance around the structure, with application rate tied to a 5- or 10-foot band width. Hand-operated granular applicators are best suited for this application, and some are calibrated by the manufacturer.

- *Boundary application* refers to application to soil, ornamental gardens, vegetation areas, and areas covered with a ground cover of organic mulch.

- *Broadcast application* is intended to thoroughly and uniformly cover the surface of the treated area. Application is usually made with a calibrated mechanical spreader which is either hand pushed or equipment drawn.

- *Spot application* is usually for the control of actively foraging ants and may involve mechanical spreaders.

Traps. Liquid food lures are used for controlling nearby nests of yellowjacket and umbrella wasps. Liquid attractant is placed in hanging traps. The trap design facilitates the entry of individual wasps but not their exit. The strategy is to deplete the colony of workers, which deprives the larvae of food, and eventually causes the colony to decline and die out.

ENVIRONMENTAL CONDITIONS

Conditions during and after outdoor application can influence the residual deposit of insecticide, and, ultimately, the control achieved. Immediately following application, environmental conditions begin to degrade the residue. The loss of insecticidal activity begins with evaporation of the water carrier and some of the active ingredient, followed by degradation by ultraviolet light, abrasion of the surface, and absorption into the surface. Much of the loss of activity is due to movement of the insecticide into porous surfaces.

Ultraviolet Light. Nearly all insecticides are affected by long-term exposure to sunlight. The ultraviolet (UV) light component of sunlight can degrade the chemical structure of materials. However, the formulations of outdoor-use chemicals are usually designed to protect the active ingredient from UV degradation.

Moisture. Rainfall following application will remove insecticide residue on non-porous surfaces or increase absorption into porous surfaces, and generally reduce toxicity. Treating wet surfaces increases the potential for run-off and/or decreases residual activity.

However, rainfall following application of granular formulations for soil-insect control can benefit efficacy by increasing the release of insecticide from the inert carrier.

Air Temperature. In general, air temperature has little affect on spray treatments, except to help decrease the drying time for treated surfaces. High ambient air temperature can increase insect metabolic activity, and this may enhance the effect of the insecticide once it enters the body.

Temperature inversion, that is, a layer of cold air close to the ground, occurs early in the morning, after the ground temperature has fallen during the night. Under these conditions, spray droplets tend to remain close to the ground, reducing drift.

During the hottest part of the day, small spray droplets drift upward in warm air, rather than drifting horizontally. This may position droplets to be drawn into buildings through open windows or ventilation systems.

Air Currents. Not all the droplets leaving the sprayer nozzle contact the target substrate or pest. Small droplets created at the edge of the spray pattern can be carried away by air currents, and large droplets may bounce and fragment into small droplets when they hit the surface.

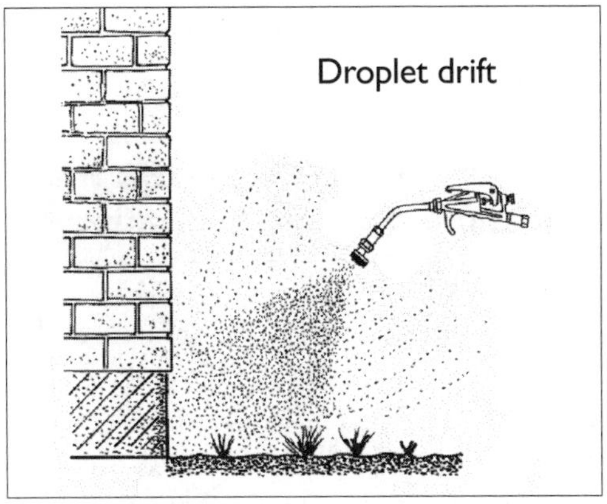

In air currents as low as 4 mph, small droplets will drift downwind. Although these droplets carry only a small amount of insecticide, they can affect fish ponds and flowering plants, which further impact honey bees.

Spraying when surface winds are above 9.5 mph will likely result in insecticide drift of large droplets to non-target surfaces.

SURFACES AND SUBSTRATES

Perimeter treatment involves liquid or granular application to the foundation wall, siding, door and window framing, and soil adjacent to the wall. Surfaces treated at these sites have features and conditions that influence the insecticide residue. Exposure to seasonal rainfall and sunlight will limit the residual activity. Although the east-facing side of the house receives the same sunlight as the one facing west, the west-facing side is exposed during the hottest part of the day, making it the warmest. West-facing surfaces provide insects with warm conditions and limit the residual life of insecticides.

The south side of a building is heated more in winter than in summer. This side may show the first signs of insect activity in spring and require first attention when making a perimeter application.

Porous Surfaces. Most outside substrates have a rough or uneven surface and are considered porous. The most common of these are concrete, stucco, brick, acrylic-painted, and bare wood. Porous surfaces generally decrease the efficacy of insecticide formulations. This is due primarily to the insecticide being adsorbed into the substrate matrix and becoming unavailable for contact by the pest.

Microencapsulated and wettable-powder formulations generally provide greater exposure to pests than emulsifiable concentrates. However, some microcapsules may become lodged within the matrix of the substrate and not be available for pick up or crushing by the pest.

- *Residue.* Porous substrates influence the residual activity of insecticides. The degree of influence depends on the surface, formulation of the insecticide, and exposure. Porous surfaces are sometimes characterized by an alkaline pH which favors a rapid degradation of insecticide. Emulsion concentrates experience more rapid breakdown on exposed surfaces than wettable powders, which are held on the surface and protected by the carrier in the formulation.

Insecticide labels may permit increased concentrations or application volumes, and more frequent applications when treating concrete.

- *Wood.* Bare-wood substrates include shake siding and roofing, attached decks, landscape timbers, and fences. The surface of weathered wood is porous, and insecticide labels may permit higher application volumes to achieve the desired coverage. Solvent- and water-based formulations will penetrate wood but remain in the outer layer. These residues provide protection and control of insects that tunnel into the wood. Wettable powders and microencapsulated formulations can provide some surface residual activity, even against wood-infesting insects.

In the case of borates, frequent exposure to rainfall can improve penetration, rather than reducing it through leaching.

Non-Porous Surfaces. Galvanized and stainless steel, aluminum and vinyl siding are common non-porous substrates. Stainless steel may accelerate the degradation of certain insecticides and limit their efficacy. Aluminum is an inert substrate, but is subject to rapid heating, which may limit the residual efficacy of insecticides. Painted wood surfaces are porous, and they limit the activity of insecticides: rates of insecticide uptake from painted wood by crawling insects may be only 1% to 2% of the amount of transfer that occurs on non-porous substrates.

Vinyl siding is inert, but, as it ages, a small amount of hydrochloric acid may be released onto the outer surface, which can affect insecticide residue.

LANDSCAPE FEATURES: RESIDENTIAL

Fences. A common feature of fences is setting of the support posts deep into the ground. However, this exposes the bottom of the post to wood decay fungi and soil organisms. Frost can damage or unsettle fence posts in soil with poor drainage. When water in the soil around posts freezes and expands, the posts may be pushed up from the base. Water that freezes and expands inside wood posts will crack the concrete around them. Both events can result in exposure of the post to ground moisture, which favors wood decay organisms.

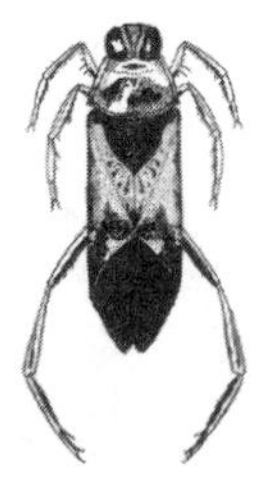

Backswimmer

Swimming Pools. The insects and other arthropods associated with swimming pools are accidental invaders. There are few aquatic insects that use regularly maintained pools as a breeding site. Wind-blown insects from surrounding shrubs and trees include aphids, leafhoppers, and crickets. Insects attracted to outdoor lights may accidentally or intentionally move to the pool, including species of aquatic diving beetles, water striders, backswimmers, water boatman, and giant water bugs. Some of these may remain in the water for extended periods and can deliver a painful bite. In addition, crawling invertebrates from the surrounding groundcover may fall into the water, including wolf spiders, millipedes, springtails, and ground beetles.

When in the vicinity of pools, applicators of liquid or granular insecticide to vegetation and perimeter groundcover must guard against drift. Spray droplets from coarse fan nozzles on tank and backpack sprayers can deposit in pools that are within 10 feet of application.

Utility Meters and Heat Pumps. Electric-service meters and telephone- and cable-service boxes are located on the outside of houses. They are small structures and usually placed in an exposed site to make them easy to locate and service. Insects that nest in these structures include umbrella wasps, yellowjackets, fire ants, and spiders, including the black widow spider. The fan and compressor of heat pumps may discourage wasps from nesting, but they will not deter spiders from making webs around the outside. Liquid pesticide application to these sites could contaminate motors and electric circuits. Dust formulations may be effective but more difficult to apply.

Termiticide soil rodding to perimeter soil must consider buried cables adjacent to these utility boxes.

Garden Pools and Fish Ponds. Landscape pools and fish ponds within 20 feet of buildings should be considered when spraying insecticides for perimeter pests. Label directions typically stipulate applications be made toward the foundation wall and adjacent soil. Droplets from spray applications may be carried off-target by 3-mph air currents. Fish raised in ornamental ponds are generally sensitive to insecticides. Caution should be used when applying granular insecticides to ground cover close to ornamental pools.

Droplets from a coarse fan spray can drift 10 feet from the spray site. Power mist droplets from a backpack sprayer can drift more than 150 feet.

Animal Pens. Outdoor enclosures or fenced yards for household pets may be used seasonally or year round. These sites have animal shelters, food and water containers, areas of exposed soil, and animal excrement. Pet food may be continuously available. Pest insects and other arthropods in this habitat include mosquitoes, flies, fleas, and ticks. Various insects and rodents are associated with the pet food when it is available outdoors.

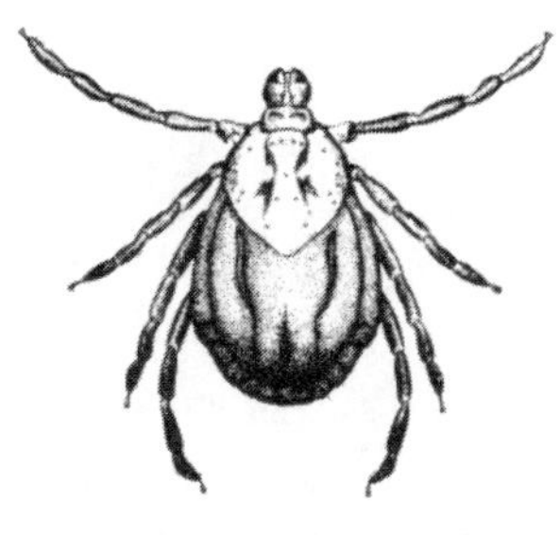

American dog tick

Insecticide labels recommend domestic pets be kept off treated surfaces until the substrate is completely dry. Pet food and water containers within 20 feet of spray applications may be contaminated with insecticide drift.

Bird Feeders and Baths. Bird seed scattered from feeders may serve as a food source for rodent pests that occur around the perimeter and inside buildings. Birdseed mixes contain seeds of millet, sunflower, safflower, and rapeseed, all of which are fed on by mice and voles when found scattered on the ground. White-footed mice and deer mice invade buildings at the ground level and these small rodents utilize seeds scattered under bird feeders.

Bird feeders are usually close to the house perimeter. Although feeding stations may not be in the path of perimeter insecticide application, the seeds scattered on the ground and fed on by some bird species may be accidentally exposed to liquid or granular insecticide. Bird baths may be elevated several feet or at ground level and concealed. In either position, they may be exposed to insecticide drift during perimeter application.

White-footed mouse

Trees, Stumps, and Firewood. Live trees provide food, harborage, and nest sites for a large number of insects. Aphids and other honeydew-producing insects may become abundant in landscape trees, and these populations can support ant species, including carpenter ants. Subterranean termites often move in the cavities around tree roots in the ground around buildings. Tree holes, cavities, and dense canopies can provide environmental conditions suitable for outdoor populations of the smokybrown cockroach.

Dead or damaged portions of tree trunks, roots, and abandoned stumps can be used by subterranean termites and wood-infesting ants as food or nest sites. The root system remaining after the uppermost part of the stump is removed will decay in several years, but can provide food and foraging channels for subterranean termites. Pine and hardwood logs can provide harborage for pests that occur indoors, including wood cockroaches, earwigs, scorpions, and crickets. Logs can also be infested with longhorned beetles, carpenter ants, and termites, depending on the level of moisture and decay.

Applying insecticide to the surface of firewood logs will result in little or no control of any pests, and it is usually not permitted by insecticide labels.

Vegetable Plants and Gardens. Small-plot vegetable and herb gardens may occur in yards, along with plants in containers placed close to houses, and those grown around the perimeter. These plants are identifiable by their late-growing fruit, but prior to that, they are not easily distinguished. Pesticide application near vegetable and herb plots may result in drift and some level of contact with the leaves and fruit.

Liquid termiticides are applied to a 6-inch trench along the building perimeter, and into the soil to a depth of about 4 feet along the foundation wall above the footing. Following termiticide application in spring, there may be concern about translocation of insecticide into vegetables planted in this soil. The label directions for some termiticides prohibit treatment around edible plants.

Pesticides may damage vegetables when used improperly or too closely. Damage usually occurs within 24 hours of when the plants were sprayed or exposed to pesticide drift. Damage from drift of liquid insecticides often occurs when daytime temperatures are above 90°F. It may also occur during damp, humid weather when spray droplets are slow to dry on leaves.

Pesticides applied from tanks that previously held herbicides may kill or burn leaves of plants in the spray or drift path. Traces of herbicides are difficult to remove from the tank, pump mechanism, hose, and sprayer valves.

Insect, mite, and fungi damage to vegetables and landscape flowers may be attributed to accidental spraying or soil contamination. Fungi-attacking plants can cause wilting, yellowing, and spotty leaves. A white, powdery mildew deposit on leaves is caused by fungi, but may be mistaken for insecticide deposit. Spider mites damage vegetables and flowers by sucking plant sap and removing the green pigmentation in leaves, while mite infestations can produce yellowing along the edges of leaves. Flea beetles chew small holes in the leaves of garden plants; these small beetles cause damage during spring and summer.

LANDSCAPE FEATURES: COMMERCIAL

Temperature. The perimeter zone around the foundation of commercial buildings is usually warmer than it is further away from the wall. This is due to heat gained by the foundation, sidewalk, or other heat-absorbing material. Boxelder bugs, Asian stink bugs, and cluster flies actively select sun-heated surfaces on buildings for mass aggregations in the fall. Boxelder bugs are capable of detecting temperature differences as low as 1 degree, and moving to those surfaces and then indoors. Application of insecticides to the warm surfaces late in the day may extend the residual activity of the application and provide control into the night. Application around heat pumps and air-intake vents must be avoided.

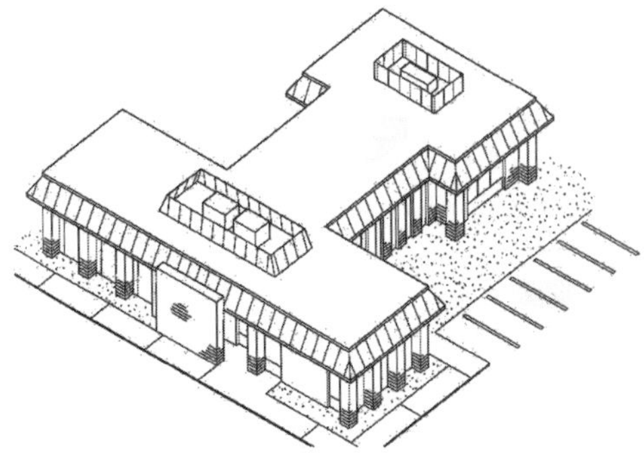

Perimeter zones

Vending Machines. These are often an add-on feature to a store or service center and can be located near an entry door. Modern vending machines have bright lights, electric motors, and refrigeration compressors that generate heat. These features, and the presence of food materials, attract insects and rats. Rodent bait stations and/or snap trap stations can be secured behind vending machines.

Garbage Stalls. Restaurants, apartment buildings, schools, and other commercial facilities usually have a separate stall for dumpsters. Fast-food restaurants may also have separate containers for waste cooking oil. The organic matter and clutter in these stalls provides food and harborage for house flies, fruit flies, and rats. Typically, the distance to the main building is short, and the presence of food odors is attractive. Rat- and fly-control programs for restaurants must include the garbage dumpsters on the property and in adjacent properties.

Outdoor Lights. Lighting and lighted vending machines contribute to the presence of insects around commercial buildings. Termite swarmers, flies, beetles, stink bugs, and moths are attracted to bright lights at night. This behavior often results in insects coming indoors and dead or dying insects near the light source.

Mercury vapor lamps are often used for outdoor lighting. There is a strong ultraviolet and blue light content to these bulbs, and they provide a strong attraction to insects at night. Sodium-vapor lamps are an alternative to mercury-vapor lamps and attract few insects. They have a distinct yellow light because they produce almost entirely one wavelength of yellow light and only a small amount of ultraviolet light.

CHAPTER 16

RODENT CONTROL

Rodent control is the most challenging service of professional pest control. The Norway rat, roof rat, and house mouse have been associated with the human environment for thousands of years. They are successful pests because they can eat our food, observe and remember objects and events, learn from past experiences, and quickly adapt to changing conditions. Unlike insects, which are short-lived and guided primarily by instinct, rodents have a brain that stores and uses information. Rodents live as individuals, and each has a unique set of behavior patterns developed over its lifetime.

DISEASE TRANSMISSION

The foraging and movement habits of rodents can expose them to pathogenic organisms, and under certain conditions, they can transfer these organisms to people or indoor surfaces. Rats occur in sewers and garbage where they can contact *Salmonella* and later spread these bacteria in their feces. Mice can spread *Salmonella* in poultry operations. Although the plague bacterium does not occur in commensal rats in the U.S., it is in wild populations of ground squirrels and other rodents in western states.

Hantaviruses. This virus is spread in the urine, fecal droppings, and saliva of rodents, particularly mice. The common carriers are deer mice, cotton rats, and rice rats in the Southeast, and the white-footed mouse in the Northeast. The deer mouse is considered the primary carrier of the virus. States reporting the most number of Hantavirus cases are Arizona, California, Colorado, Idaho, Montana, New Mexico, and Utah.

Norway and roof rats and the house mouse are not carriers of Hantavirus.

Hantavirus Pulmonary Syndrome is spread to humans when dried material contaminated by rodent urine or feces is disturbed. This can occur through cleaning of nest sites or traps, or handling of dead animals. Humans become infected by inhaling aerosol-sized particles contaminated by the virus. The virus is rare, but individuals involved in rodent control programs should take precautions, especially when removing deer mice from traps or handling bait stations contaminated by rodent urine or feces.

For routine house mouse control programs, the use of a respirator may not be necessary. Wearing gloves when working with traps or bait stations can provide protection from pathogenic organisms and rodent parasites.

STRUCTURAL DAMAGE

Rats and mice gnaw holes in surfaces to gain entry into buildings and into food containers; this damage can be extensive when infestations are large. The hard enamel on rat incisors enables them to gnaw through aluminum, copper, and iron, and some building materials. The incisors are separated from the molar chewing teeth by a gap, which allows rats to close their mouths behind the incisors while gnawing. Thus, they can cut through material that is inedible or distasteful without eating it.

The teeth of a rat have a hardness index of 5.5, which is greater than iron (4.0), copper (2.5), and aluminum (2.0).

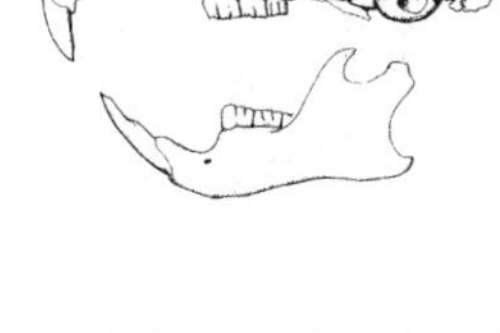

Rat skull

A rat can gnaw and enlarge any 1/2-inch opening into an entry point.

IDENTIFICATION

Norway rat and feces

Norway Rat *(Rattus norvegicus)*. This species is also known as the brown rat, sewer rat, or wharf rat. It is larger than the roof rat.

- *Identification.* The head and body length is 7.5 to 10 inches, and the tail is 6.0 to 8.5 inches long. The adult weighs 0.75 to 1.25 pounds. Fur is coarse and brown, with gray to yellowish white underside. The tail is dark above and light below; ears are small.

- *Fecal droppings* are 3/4-inch long and capsule shaped, both ends are usually blunt; they often contain hairs.

Norway rats are distinguished from roof rats by a blunt nose, small eyes and ears, and a tail that is shorter than the head and body length.

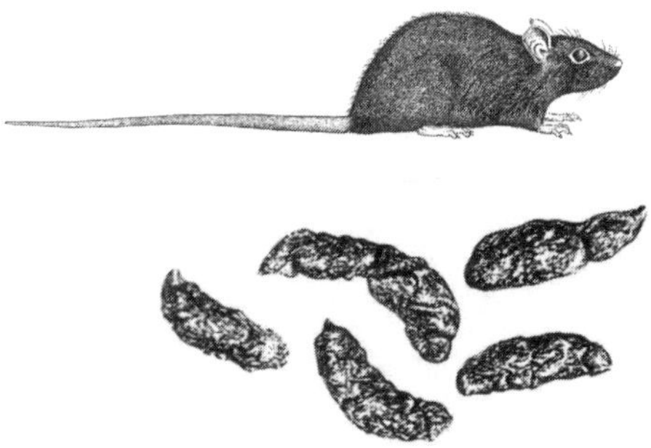

Roof rat and feces

Roof Rat *(Rattus rattus)*. This species is also known as the black rat, ship rat, or palm rat.

- *Identification.* Smaller than the Norway rat, its head and body length is 6.0 to 8.5 inches, and the tail is 7.0 to 10.0 inches long. The tail is generally longer than the combined head and body length. The adult weighs 0.5 to 0.6 pounds. The tail is uniformly dark in color; the ears are large and prominent.

- *Fecal droppings* are 1/2-inch long and spindle-shaped, with one end usually pointed; they often contain hairs.

House Mouse *(Mus musculus)*.

- *Identification.* This mouse is dark gray above and light gray on the belly, and the tail is lighter below than above. The head and body length is 2.5 to 3.5 inches, and the tail is 3.0 to 4.0 inches long.

- *Fecal droppings* are 1/4-inch long and rod shaped.

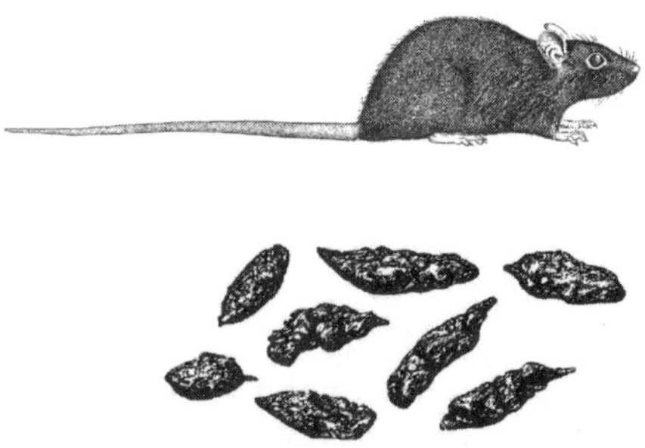

House mouse and feces

Adult mice look similar to young roof rats, but young rats stay close to their burrows, so would not be found moving about.

Deer Mouse *(Peromyscus maniculata)*. This species is also known as the field mouse.

- *Identification.* The body is reddish brown above, with a white chest and white feet. The head and body length is 5.0 to 8.5 inches, and the tail is 2.0 to 4.5 inches long. This is the most common and widely distributed mouse in the U.S., but it is absent from parts of the eastern coastal and Gulf states.

- *Fecal droppings* are 1/4-inch long and elongated.

White-Footed Mouse *(Peromyscus leucopus).* This species is also known as the wood mouse.

- *Identification.* The body is grayish to dull brown above, with a white chest and white feet. The head and body length is 6.0 to 8.0 inches, and the tail is 2.5 to 3.5 inches long. This mouse is found east of the Rocky Mountains, but it is absent from coastal regions of the eastern and Gulf states and Florida.
- *Fecal droppings* are similar to those of the house mouse.

White-footed mouse

Deer mice and white-footed mice do not often occur in urban areas but are common in rural areas. They are active year-round.

Voles, Meadow Mice *(Microtus pennsylvanicus).* This species is also known as the "field mouse," but the appearance and habits of voles are very different.

- *Identification.* The body is blackish brown above and gray below. The head and body length is 5.5 to 8.0 inches, the tail is short, about 2 inches long.
- *Fecal droppings* are elongate and dark colored, little different from mice.

Vole (Meadow mouse)

Voles are found in the Midwest and Eastern U.S., especially along the coast, south to Florida. Voles make extensive runways below the soil surface; after snow melts, their above-ground tunnels can be seen. Above ground, they eat plant seeds, underground tubers, grass, and clover.

Moles *(Parascalops breweri, Scalpus aquaticus, Condylura cristata).* There are several species that occur in the Eastern U.S. The scientific name of the Eastern mole, *Scalpus aquatica*, refers to its webbed toes; it is not aquatic.

- *Identification.* The hairy-tailed mole is 5 to 7 inches long, the Eastern mole is larger and has a short, hairless tail. There are no visible eyes.

Hairy-tailed mole

Moles spend most of their time below ground; they eat earthworms and insects found below the soil surface. Their underground habitat limits the predators that attack them, and populations can become large. They have one litter a year with 2 to 6 young.

Hairy-tailed moles are active above ground during the day and are often caught by house cats.

INSPECTION SIGNS

Rodents are usually active at night and are very wary, so they are rarely seen during inspections.

Burrows. Norway rats make ground burrows that are about 4 inches in diameter, with one or more entry points. They may be located anywhere in dense vegetation, along the edge of concrete slabs and foundation walls, and adjacent to dumpsters and garbage collection sites.

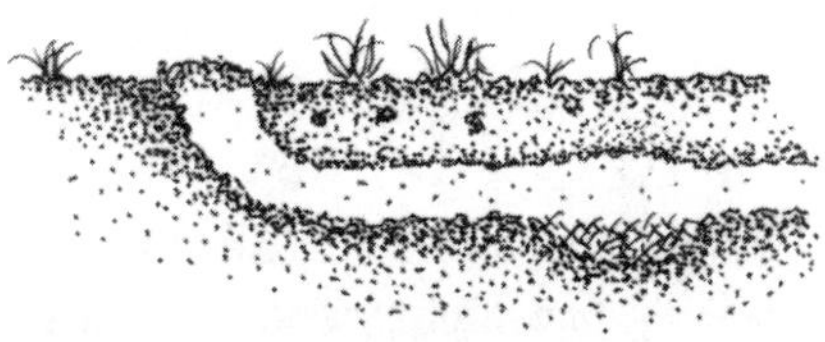

Ground burrow

Roof rats burrow only occasionally. When they do, the burrow entrance is surrounded by a small mound of excavated soil. The system may have escape holes, which contain loosely packed soil that can be easily pushed out. Entrances that are closed during the day may be reopened at night.

Entry Points. Locating entry points to a structure requires inspection of all possible access points. Rats and mice will gnaw to enlarge a small opening to make it suitable for them to pass through.

Young rats can enter openings 1/2 inch in diameter; young mice can enter openings about 1/4 inch in diameter. Holes used by adult rats are 3 inches; holes used by mice are about 3/4 inch.

Fecal Droppings. Fecal droppings are found along movement paths and wherever rats and mice are active. Accumulation of droppings can indicate where rats and mice spend considerable time, and these may be good locations for the placement of bait stations or traps.

A single adult Norway rat produces about 37 droppings per day, an adult roof rat produces about 50, and a house mouse produces at least 50.

The number of droppings may be expected to help provide an estimate of the size of an infestation, however the number is also dependent on the type of food, moisture content, availability of water, and rodents eating fecal pellets.

Gnaw Marks. Rats will gnaw on almost anything they can get between their incisors, and they have a bite force of about 1-pound pressure.

Nest Sites. Nests and food caches may be indoors when there is an ongoing infestation of rats and mice. Shredded paper, cardboard, cloth, furniture stuffing, and insulation may all be used for nesting.

- *Norway rats* prefer to nest near the ground, such as in the lower stories of buildings; nest sites include under cabinets, around or in undisturbed boxes, and in stuffed furniture. Nests are about 8 inches in diameter.
- *Roof rats* prefer to nest above ground, along rafters and in attics.
- *House mice* nest where there are undisturbed conditions, including in wall voids and furniture, and under cabinets. Nests are about 5 inches in diameter.

Sounds. Rats and mice make a characteristic set of sounds, especially at night when moving in wall voids, attics, drop ceilings, and closed spaces. The sound of individuals gnawing on household objects can also be heard.

Tracks and Runs. Rat footprints or marks made by its tail can be seen on dusty surfaces. The tracks of the five-toed rear feet are more noticeable than those of the four-toed front feet. In addition, the greatest body weight is over the hind feet.

Ultrasonic sounds produced by mice are in the 70 kHz range. This is beyond the upper limit of hearing of the domestic cat, which is 64 kHz. Cats are more effective as visual rather than sound-oriented hunters.

Tail marks are a broad line, along with footprints in a dusty surface. These are more likely to be made by adult Norway rats, since roof rats carry their tails high when walking.

Rub Marks. Rub marks consist of natural oils from the rat's fur. These are deposited when the animal rubs against a wall or corner when running. Roof rat rub marks are usually found in elevated locations. They may produce a "halo" rub mark, in which a dark, curved band of grease replicates the outer edge of the rat body.

Urine Stains. Rats and mice dribble urine whenever they move; urine spots can be found along frequently used runways and on food material. Whether wet or dry, urine stains of rats and mice fluoresce when exposed to ultraviolet (UV) light. Searching for urine stains with UV light can illuminate rodent activity and building entry points.

Dried urine fluoresces with a bluish white light if relatively fresh, but yellowish white if old.

MOVEMENT AND BIOLOGICAL SENSES

Rats and mice are nocturnal animals, but may be out during the day if there is no human activity, or the location is well concealed. There are two periods of adult peak activity: a short time after dark and an hour before dawn. However, the need for food and mating can change these activity periods. Once rodents leave their nest sites, they select pathways with familiar features and scents. These routes provide concealment and best escape, and are the shortest distance to food.

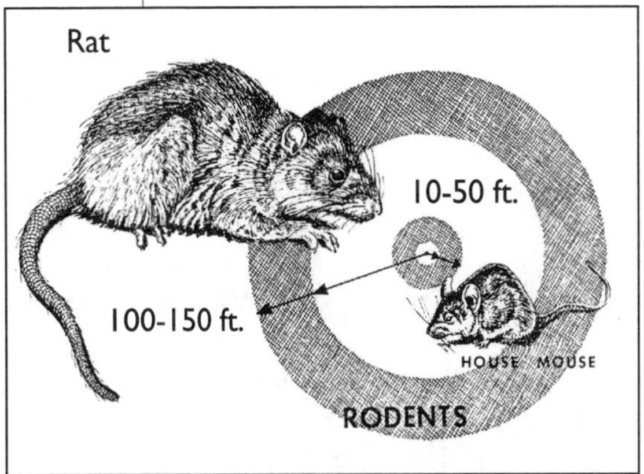

Movement/Foraging Range. The movement range of rats and mice is the area that includes the network of routinely used paths, runways, and escape routes. The size of the area depends on the location of food in relation to shelter and safety. The whole of the movement range may not be utilized at the same time, especially if food and shelter remain stable. The movement range tends to become large when food and shelter become scarce or with frequent structural changes.

Rats. Norway rats have a movement range of 100 to 150 feet in diameter; greater diameters may occur when the distance to food increases. Roof rats have a large and three-dimensional range because of their climbing habit.

Mice. House mice have a 10- to 50-foot-diameter movement range, but it may be only 6 to 10 feet when food and safe harborage coincide. The movement range may increase to include moving between buildings.

Biological Senses. All movement and foraging activity is influenced by information received by the biological senses of hearing, vision, smell, taste, and touch.

- *Hearing.* Rats and mice can hear and respond to humans talking, dogs barking, or sudden noises.

- *Sight.* Rats and mice have poor daylight vision and do not see colors, but they have good night vision. Their eyes can detect movement in dim light, and they can recognize patterns and objects.

Rodents can detect movement at a distance of about 30 feet.

Although house mice infesting buildings form separate breeding units, individual mice may move between buildings. Buildings may be re-infested from adjacent buildings in which control was not achieved.

Smell. Rats and mice continually move their heads while sniffing, and they often rise up on their back legs to sample the air. Rodents leave odor trails of urine and other secretions to mark pathways and territories. The odor of humans is not a deterrent to their activity and movement.

Taste. Rodents nibble and taste substances they encounter. The Norway rat feeds on one food at one time, while the roof rat and house mouse are erratic feeders. Rats can detect impurities in food as low as 1 ppm, and they may reject food associated with a previous illness.

Touch. Their well-developed sense of touch enables rats and mice to run along regular pathways, as well as in unfamiliar areas. The long, bristled vibrissae extend laterally from the face, and the "guard" hairs extend from the fur on the back provide guidance. This "constant-contact" behavior causes the rub marks along runways.

EXPLORATION AND AVOIDANCE

Day and night movement by rats and mice is conducted by memory. They memorize the features of runways and hiding places, and the location of food. They travel the same routes every day or night, so that the muscular movement required to take the path to food and back is learned and continuously reinforced.

When a regularly used pathway is blocked or changed, rats will become disoriented but try to continue their travel. The interruption may last a few minutes or hours before they reorient to the feature or obstacle.

Exploration. When rats and mice enter a new area, the primary behavior is exploration. The objective is to gather information on the location of food, shelter, and escape routes. This behavior dominates their activity until they become familiar with the environment and establish safe and familiar pathways to resources.

Traps and bait stations placed along perimeter walls and at entrance doors and openings take advantage of rodent exploratory behavior.

Neophobia. Neophobia, the fear of novelty, is characterized by the avoidance of any new object, especially along a pathway or regularly used route. Neophobia is common in Norway rats and roof rats, and it is triggered by any change, including placement of a new object in their environment or a familiar object in a new position. This behavior is based on the contrast between what a rat expects, based on experience, and what it encounters. Neophobia impacts the use of rodent traps or bait stations, because the traps or stations must be placed where rats or mice will encounter them, but this may also trigger avoidance and diminish effectiveness.

Neophobia lasts a few hours to days before the rat(s) accept a new object into their regular routine.

Placing unbaited or baited but unset traps for a few days can help to reduce the potential for avoidance.

Mice may increase exploration when their habitat is disturbed, which can increase the efficacy of traps or bait stations. Mice will feed on recently placed bait even when familiar food is available.

BIOLOGICAL ABILITIES AND HABITS

Preventing access to the inside of buildings, and eliminating rats and mice once they are inside, requires an understanding of their biological abilities and feeding habits.

Burrowing. Norway rats seek harborage or entry to buildings by burrowing around the foundation. They begin burrowing a short distance from the edge and dig down at a steep angle until they contact the foundation wall. Roof rats sometimes burrow beneath rocks and along foundation walls, and they will invade sewers.

Reaching, Jumping, Climbing. Rats can jump about 24 inches and mice at least 12 inches. Norway rats can leap about 36 inches and mice about 26 inches. Both can drop 50 feet to the ground without being hurt. Rats and mice can climb anything they can grip with their feet and claws, including vertical pipes up to 4 inches in diameter. Smooth surfaces can be climbed if there is an adjacent support. Norway rats have less developed climbing ability.

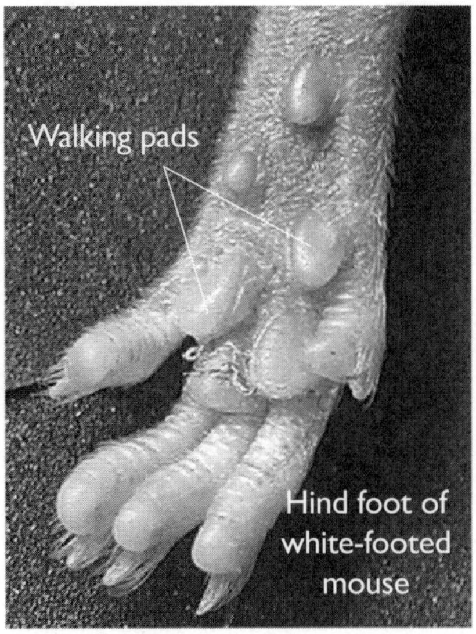

Swimming. Rats are good swimmers; they can tread water for hours and swim long distances. Norway and roof rats may enter buildings through the water trap of a toilet bowl and the floor drain connected to a sewer line. Although mice can swim well, they rarely use this ability to enter buildings.

Food Preferences. Rats and mice will feed on almost any packaged or fresh human food. Municipal garbage is nutritionally sound, so urban rats mature rapidly and have a large body size. Parental food preferences of rats are passed to the young in the mother's milk and food residues on her face and legs.

House mice eat about 1 ounce of food per day; they prefer cereal grains and seeds. Mice nibble and sample a number of food items while foraging.

Norway rats eat about 9 ounces of food per day. They tend to remain loyal to a few food sources.

Water is an attractant for rats and mice if it is within their foraging territory. Mice are able to survive without free water, relying on that which can be obtained or produced from food digestion. Norway and roof rats may go for long periods without drinking if their food has high moisture content.

Reproduction. In a favorable habitat, rodents can increase rapidly because of their short gestation period, large litter size, rapid maturation, and frequent breeding.

- *House mouse:* Female is reproductively mature within 8 weeks; she can have 10 litters and produce 30 to 60 young per year. Young leave the nest within 30 days. Female life span is about 1 year.

- *Deer mouse:* Female is reproductively mature within 6 weeks; she can have 11 litters and produce 40 to 55 young per year. This mouse breeds year round, but most often in the warm months. The young leave the nest within 25 days. Female life span is about 1 year.

- *White-footed mouse:* Female is reproductively mature within 7 weeks; she can have 8 litters and produce 30 to 40 young per year. Breeding season in northern regions peaks in spring and late summer, in southern regions breeding is year-round. Young leave the nest within 25 days. Female life span is about 1 year.

- *Norway rat:* Female is reproductively mature by 3 months; she can have 7 litters and produce 19 to 56 young per year. Young leave the nest within 30 days. Female life span is about 1 year.

- *Roof rat:* Female is reproductively mature by 3 months; she can have 9 litters and produce 12 to 56 young per year. Young leave the nest within 30 days. Female life span is about 1 year.

RODENT CONTROL

Rodent control for commercial facilities is based on a long-term program that includes strategies for indoor elimination and prevention. At this level, emphasis is placed on regular inspection and perimeter baiting or trapping to achieve and maintain a rodent-free environment. At the residential level, rodent control usually has the short-term objective of eliminating the pest(s) indoors, but should include exclusion methods to prevent re-infestation.

The principal of "situational specific control" includes the concept that each site has unique features or constraints, and control strategies must be designed for these conditions. The basic components of rodent control are: inspection, sanitation, exclusion, and control. The implementation of these depends on the location, the pest, and the level of infestation. Successful programs are based on site-specific strategies designed by a professional.

Inspection. Rodent control programs rely on a thorough inspection. The objectives are to determine the harborages and foraging areas of the infesting rats or mice, and identify the physical and environmental features that contribute to the infestation.

Sanitation. Rodent infestations are limited by the availability of food and shelter. Sanitation methods decrease available food, remove harborage and nesting sites, and disrupt pathways to food or shelter.

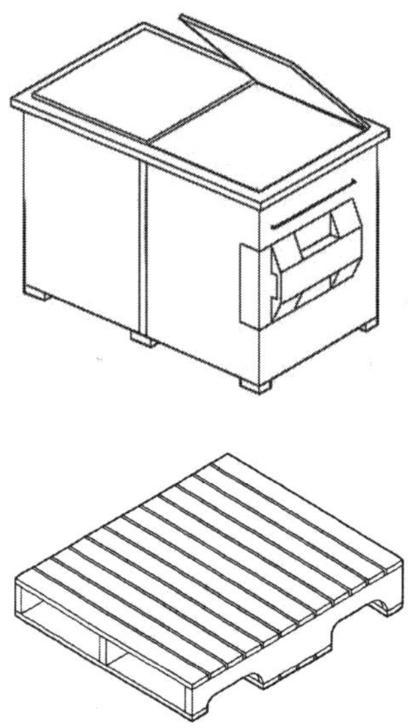

Potential rodent harborage

- *Dumpsters* are often overlooked as breeding or feeding sites for rodents infesting nearby buildings. These containers are usually coated with or surrounded by grease and food debris, and the drain holes at the bottom are often open or the plugs are missing. Dumpsters are usually placed close to the back door of food facilities. Thus, the distance to the building is usually short, making it easy for rats to find food, water, and shelter.

- *Wood pallets* provide ideal harborage for rats and mice. Indoors, pallets are usually not inspected for signs of rodent activity. Regular replacement can discourage their use as shelters, but moving infested pallets can transfer the problem to other sites. Outdoors, pallets are often stacked and may remain so for long periods. These provide ideal harborage and give access to the adjacent building.

Pallets stacked 18 inches from a wall permit regular inspection from both ends and prevent rodents from establishing secure nest sites.

- *Pet food bowls* can be feeding sites for rats. Rats may become established under or near dog houses or enclosures, and take food at night. Often rats remove pieces of dog food and store them in their nests. A dog may eventually become accustomed to the presence of rats or mice in the vicinity. Indoors, mice will do the same with small pieces of dog or cat food. Rodents will also feed on undigested particles in dog feces.

Exclusion. Rats and mice enter buildings through gaps around door frames and openings along utility wires and cables. They can enter through 1/4- to 1/2-inch holes or crevices. Entry points are often difficult to locate because there may be little evidence that an opening is being used for access, especially by field mice. Rats can climb vertical pipes, including downspouts, to reach openings to upper-level floors; they can jump across gaps 48 inches wide; and they can cross narrow wires and ropes between buildings.

Baiting and Bait Stations. Most rodenticides used in commercial rodent control programs are placed in tamper-resistant bait stations. These are large profile plastic boxes with ground-level openings that rats and mice can enter. They have compartments for glue boards, devices to hold bait blocks, and/or trays to hold pellets or loose meal. The floor of the station usually has anchor ports or openings for securing to a surface, and it may be a light color to decrease daytime heating.

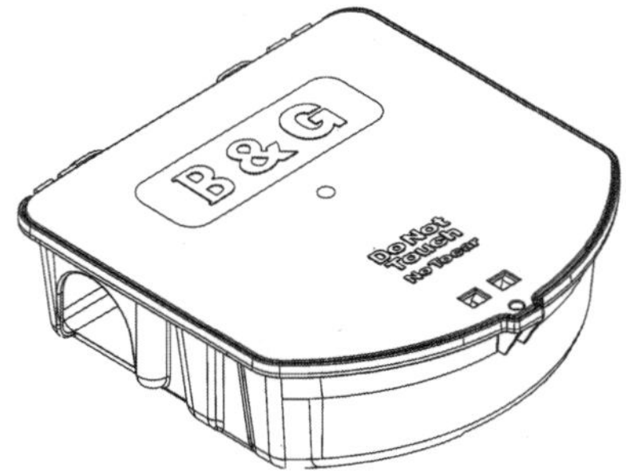

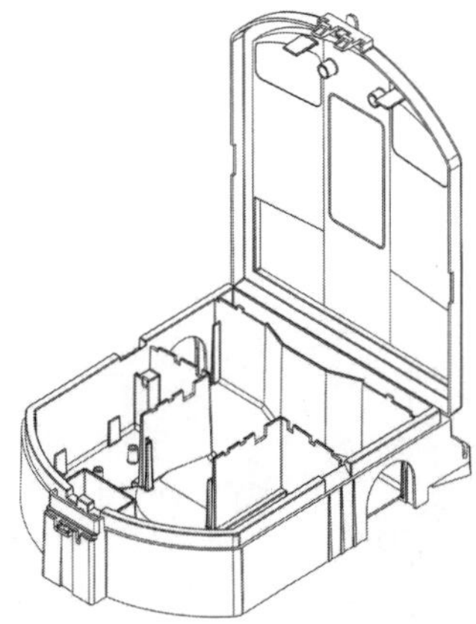

Bait stations

The benefits of using rodent bait stations are the continued availability of rodenticides to rats and security from non-target animals. The monitoring of bait consumption in a station can also enable measurement of the rodent infestation.

The design and operation of a rodent bait station is critical to a rodent control program. The station must be acceptable to foraging rats and mice, so that they will readily enter and consume a lethal quantity of bait. Station design must also consider the time and tasks needed to inspect, clean, and replenish the station. Important features include single-key opening, waterproof and hinged lid, spacing from the wall for the open lid, sturdy construction to resist damage in warehouses, and an easily cleaned interior.

- *Acceptance.* The reaction to bait and bait stations by rats ranges from what seems to be aversion and no visits to disregard and delay to attraction and immediate visits and feeding. Rat visits may be encouraged by positioning of the stations near accumulated fecal pellets, and along frequently used pathways. Removing available food or garbage in the habitat may force rodents to seek the bait blocks in the bait station.

- *Placement.* Bait stations are most effective when placed in or along rodent pathways to food. Rats generally approach food along frequently used pathways and not in open travel. Routes that lead to food influence the dominant foraging and movement behavior of adult rats; they use pathways more than landmarks as a natural map. Short paths from their burrow to food are preferred over long paths.

  Placement along a chain-link fence, for example, should have the opening to the station against the bottom edge of the fence. Rodents following a path along the fence will encounter the opening and/or detect the bait material inside. The same applies for outdoor placement against the foundation of a building, or against a wall. Select stations that allow for the lid to open against the wall without having to move the unit for inspection, cleaning, and re-baiting.

  The internal temperature of rodent bait stations may be a concern when stations are placed in sites exposed to direct sunlight and solar heating. During summer days, stations can become hot, with the temperature inside the station reaching 117.7°F in black stations on asphalt. This high temperature is brief, however, and is only slightly above the 115°F melting point of the paraffin typically used in bait blocks. The location and substrate can influence the temperature inside bait stations.

Follow fecal pellets for station placement; move some pellets into the station to encourage rats of the population to visit and feed.

Stations placed on grass in the open, such as along fences, remain cooler (91° to 94°) than stations adjacent to foundation walls. Stations on asphalt get hotter than those on bark mulch around foundations. Securing stations to concrete pavers does not influence the internal temperature.

Rodent bait stations and the substrate on which they rest retain heat after sunset—a time when rats are actively moving about and foraging. Black and gray stations exposed to afternoon sun can remain 6 to 9 degrees warmer than the air temperature 2 to 3 hours after sunset. Stations with an internal temperature of 74.8° to 79°F when the surrounding air temperature is 68.4°F may be attractive to rats that orient to temperature differences. The lingering heat in the stations may also increase or enhance the odor of the bait and attract rats to the site.

- *Anchoring.* Metal spikes can be used to secure bait stations directly to the ground; there are usually several "knock-out" holes in the base of the station that will accept a spike. The spike should be long enough to reach a firm, dry substrate below the station to meet the hold-fast requirements. Chains, wires, or plastic ties may be used to fasten stations to fence posts; stations also typically have "knock-out" slots for attaching with wires or plastic ties. Latex-based contact cement can be used to attach stations to concrete pavers or other substrates. B&G rodent bait stations can be attached to concrete pavers by a metal-clamp system that attaches to the front and back of the station.

  Physically securing a rodent bait station to a location, outdoors or in, is important to both the success of the station and the safety of non-target animals. Attaching a station to a concrete paving block, for example, can elevate it above standing water and weeds and prevent it from being moved easily. The 1- to 2-inch thickness of the paving block will not deter rats from entering the station.

EPA-established criteria defines the term "tamper-resistant," as it applies to rodenticide bait stations and securing of stations to a substrate, as: Bait stations must be capable of being anchored securely to resist efforts to move the station or displace its contents.

- *Regulations.* EPA's risk mitigation decision became effective June 2011, affecting second-generation rodenticides and the outdoor use of rodent bait stations. The regulations require that bait be placed in bait stations no further than 50 feet from a building, and all outdoor and indoor bait be placed in tamper-resistant bait stations. Loose bait, such as pellets, can not be used; only bait blocks secured inside the station are permitted.

Trapping. Trapping requires more skill and experience than baiting. Traps are most effective with small infestations or in sensitive areas where rodenticides can not be used. Traps rarely provide long-term control of an infestation, and results are best when trapping is coupled with sanitation and exclusion. When food and harborage are removed, rodents will move in search of these resources. During this time they are vulnerable to traps.

- *Preparation.* Using rat-sized bait stations in locations with both rats and mice can be effective. Stations visited by rats do not seem to deter mice from entering, and mice have increased interest and activity in the large rat stations. For rats, it is often best to start by placing unset, baited traps for a few days. This may decrease their natural neophobic response. Exposure of unset traps may also increase mouse-trapping success. If mouse-trapping results decline, traps should be moved. An increase in sanitation or habitat disruption may stimulate the remaining mice to move and forage.

Intense trapping for a few days is more effective than a few traps used over a wide area for a long time.

The presence of mouse feces in a station may deter further exploration, since the station would not be a "new" place to explore. Dumping the droppings next to the station may also discourage exploration; instead, collect and dispose of droppings away from the station placements.

- *Servicing.* Prompt removal of dead animals is important. After a few days, a rodent carcass will begin to decay and produce an objectionable odor. Rubber gloves should always be worn when removing dead rodents from traps.

Rats are not repelled by the odor or blood of a trapped rodent. Rats may, in fact, be attracted to traps that have the odor of other rats on them.

- *Humane trapping and release concept.* Some people believe that live trapping and releasing of rodents, particularly mice, is a humane method of rodent control. However, it is neither humane nor practical to capture and release rodents. Releasing captured mice outside would likely put them into the jaws of a predator within hours, or cause them to starve in an unknown habitat, or simply enable re-invasion of the original house or an adjacent house.

Use snap-trap stations that have a mechanical alert to signal when a trap has been triggered.

- *Attractants.* Food-based attractants are available for use in traps and bait stations to increase catch. They are useful in control programs in food-processing plants and schools and other locations where peanut-butter-based baits are prohibited because of potential allergic reactions. Mouse and rat attractants take advantage of the rodents' strong foraging and feeding habits.

Snap Traps. The basic snap-trap design has been improved to include a plastic or stainless steel platform, expanded trigger, baitless trigger, and stronger spring. These traps are made in sizes for rats or mice. Rat snap traps have a strong spring, and accidental release can cause harm.

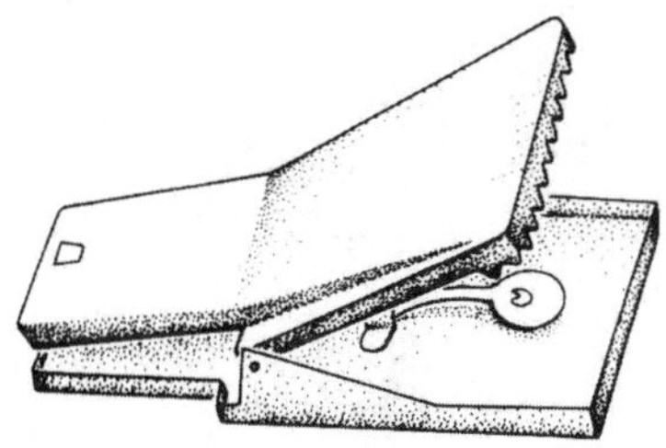

Snap traps

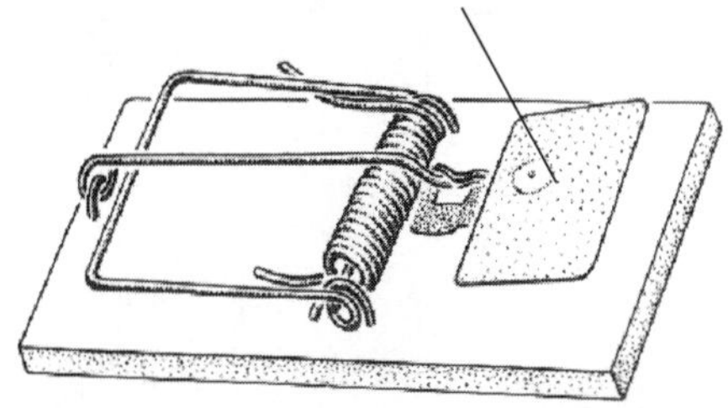

Snap traps designed for rats may not be effective for mice because mice may not be large enough to activate the trigger.

- *Bait.* Applying bait to trap triggers can increase the catch rate, but modern traps with expanded triggers can be effective even when unbaited. The commonly used baits include crunchy peanut butter, freshly fried bacon, and nuts. Use of peanut butter is usually restricted in schools, hospitals, and other sites when there is potential for human allergic reactions.

Caution: *The use of peanut butter as a bait for exposed traps has distinct disadvantages. The rodent population may not respond to this material and continue to feed on other food. This can attract ants and cockroaches to the trap site; dogs may also be attracted to peanut-butter-baited, exposed traps, since many commercial dog treats are peanut-butter based. Most important is the potential of exposure to people (and some dogs) with peanut allergies. About 1.5 million people in the U.S. suffer from a severe allergy to peanuts, and even minimal exposure can result in a life-threatening reaction for some.*

Although cheese is the traditional mouse trap bait, it is not easily attached to the trigger and not particularly attractive. For female mice, nesting material such as pieces of fabric or cotton can be better attractants.

- *Number and placement.* Snap trap success depends on placement and the use of a large number of traps. The number of traps should outnumber the suspected number of rodents 6:1. Trap placement sites include: in dark and undisturbed areas; behind appliances, furniture, and storage boxes; deep in lower cabinets and drawers that hold food or cooking pots; and along walls hidden by furniture.

- **Place traps with the trigger parallel to or against a wall or primary vertical surface, so as to be in the pathway of the rodent. Mice and rats prefer stationary objects on at least one side of them as they travel.**
- **Mouse trap placement should be at 6- to 10-foot intervals. Traps set in pairs are more effective than single traps; place traps side by side with trigger next to a wall, or end to end parallel to the wall with triggers facing away.**
- **Rat-trap placement intervals can be 10 to 15 feet. Traps may have to remain in place for a longer time for rats than mice; rats are cautious of new features in their environments and may initially avoid traps.**

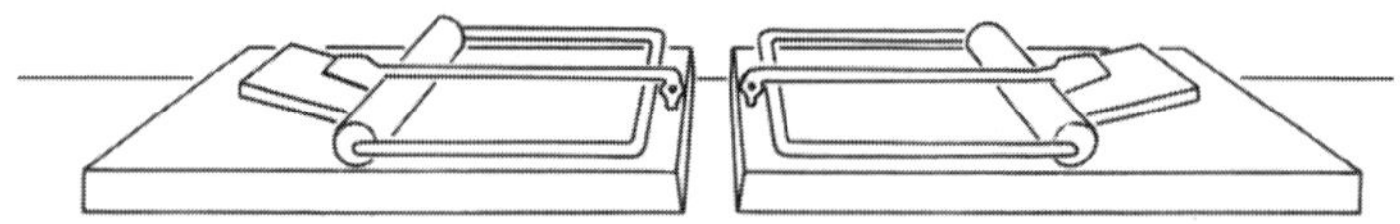

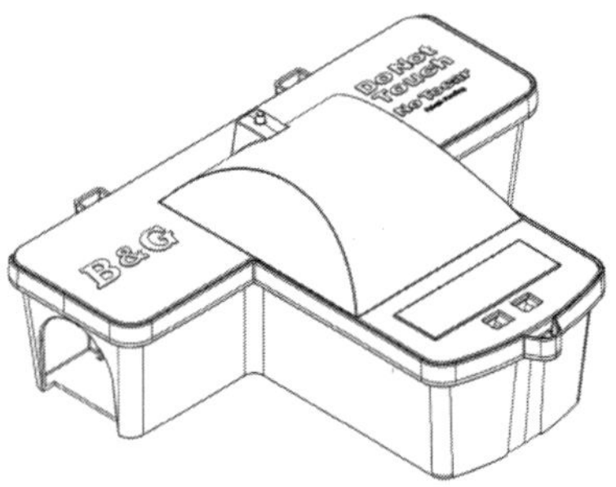

- *Snap-trap stations.* Locations that limit or prohibit the use of baits or have safety requirements for the use of snap traps will benefit from trap placement inside stations. These are small, sturdy plastic containers that are specifically designed to hold rat and/or mouse snap traps, and hide them from view. Some stations must be opened for inspection and service, but some models have an indicator that signals a sprung trap and/or captured rodent.

Multiple-Capture Traps. The benefits of multiple-capture mouse traps are their ability to operate for long periods without the use of bait and to reset themselves. High-profile, wind-up types utilize a spring-operated paddle wheel to throw captured mice into a compartment. Low-profile, non-wind-up types utilize a series of tunnels and treadles that allow mice to enter but not leave. Mice enter multiple-capture traps because they tend to investigate new holes in their territory. Captured mice then seem to attract other mice, especially littermates.

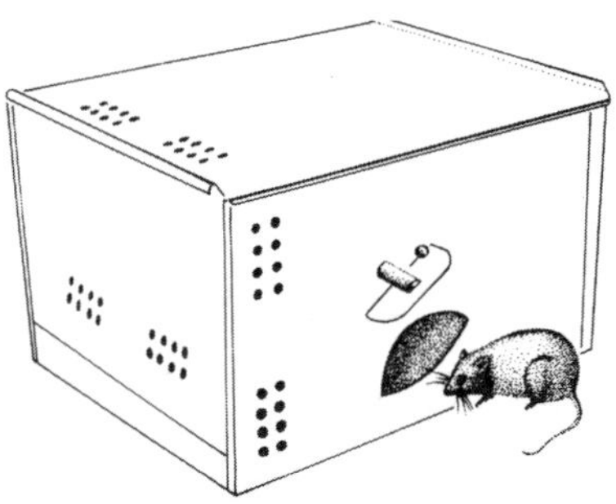

Multi-capture traps

- *Number and placement.* These traps rely on optimal placement. Pull items away from walls to encourage new routes where traps have been placed along the way. Place wind-up traps parallel to the wall, so the mice can locate the opening. Low-profile traps can be placed beneath equipment; when placed along a wall, place the entry hole against the wall. Place traps unset for a few days to improve catch.

Glue-Board Traps. Glue-board traps are used where rodenticides are prohibited. These traps have a flat, rigid base coated on one side with an adhesive that holds rodents to the surface. They are effective for adult and juvenile mice, and small and young Norway rats. Adult rats are capable of escaping if all four feet are not caught in the glue. The large traps used for rats must be secured to a surface, otherwise, partially caught individuals may drag the trap to another location.

Glue traps catch rodents alive, and individuals can remain active for a day or longer. Traps in visible locations should be serviced regularly to dispose of the captives and to make certain that rodents will not have enough time to work free of the glue.

- *Number and placement.* Effective locations are along regularly used rat pathways. The objective is to intercept the animal as it is running so that all four feet contact the glue surface. Trap efficacy may be increased by placing a small amount of bait in the center of the trap. Using large numbers of traps on the first few nights is a good strategy. Mice and rats may avoid glue-board traps, or learn to avoid them if they have escaped partial capture.

If a person or pet becomes entangled in a glue trap or glue, vegetable oil can be applied to separate the glue from skin.

RODENTICIDES

Modern rodenticides are divided into two categories: anticoagulants and non-anticoagulants. Their basic chemistry and mode of action differ, and their use by professional pest control depends on the account and the infestation.

General Information

- *Control time.* Several factors affect the time to death for rats and mice, including age and health of the rodent(s) involved and the active ingredient in the bait. Anticoagulant baits kill in about 3 days, but it can take up to 18 days.

- *Dead rat odor.* The use of baits may result in a rat dying in a wall void, sub-flooring, or ceiling space. Dead mice seldom create strong odors because of their small size. The odor of a decaying rat is strong and offensive, and it may last a week or more. As a rat carcass decays, the odor changes daily because some volatile chemicals are released early and some later in the process. At room temperature and 50% relative humidity, most of the decay is completed in 72 to 98 hours. The carcass will continue to produce an odor for several more days while chemical actions complete the decay.

- *Baiting myths.* Anticoagulant rodenticides cause death in 4 to 10 days, which gives the rodent time to get into a burrow or nest before the lethal symptoms begin. Once symptoms start the rodent rarely leaves this location.

Two myths associated with the use of rodenticides are the ability of the bait to embalm the rat so there is no odor; and the bait causing the rat to go outside to find water and die. There is no scientific basis for these beliefs.

- *Bait shyness.* Some non-anticoagulant rodenticides can produce a type of behavioral resistance called bait shyness. In this condition, the rat does not eat enough bait to be killed, but enough to become sick. The rat soon associates the taste of the bait with the sickness. The response may be to shy away from that particular bait for an extended period. Bait shyness is not a genetically-based condition and does not increase over generations. This behavior may be due to the freshness of the bait and conditions in the bait stations.

- *Secondary poisoning.* Questions about the potential for secondary poisoning often come from residential customers with house pets: If a dog or cat eats two or three mice that died as a result of eating rodenticide bait, what is the potential of the pet dying from secondary poisoning? The likelihood is very small. The dead mice have already processed and assimilated the rodenticide, and there would be little remaining in the body that is chemically active. Ingestion of bait that is still in the stomach of a mouse can occur, but the active ingredient in these few pellets is not sufficient to harm a dog or cat.

Anticoagulants. The mode of action of this group of rodenticides is twofold: they inhibit prothrombin formation, which is the material in mammal blood responsible for clotting, and they cause internal bleeding. These poisons act slowly.

- *First-generation anticoagulants.* These rodenticides include warfarin, pindone, and coumafuryl. They are considered multiple-dose or multiple-feeding rodenticides. They achieve their lethal effects with repeated feeding, because it is necessary for the chemical to accumulate in the body. Rats must feed on these baits several times over a period of 3 to 10 days; it may be longer for mice. This is because rats usually feed more intensely on baits and receive a larger dose, while mice eat small amounts at several locations and may require a longer time to ingest a lethal dose.

  These rodenticides are in ready-to-use products for consumers. The antidote, Vitamin K_1, reduces the risk of poisoning in people and pets.

- *Second-generation anticoagulants.* These rodenticides include brodifacoum, bromadiolone, and difethialone. They are single-dose rodenticides; that is, they cause mortality after a single feeding. These materials are toxic to dogs and cats. In practice, rodents feed repeatedly and can accumulate more than a lethal dose. These are more acutely toxic to rodents than the first-generation chemicals. Bait formulations contain concentrations of 0.0025% to 0.005%. Mortality occurs in 3 to 5 days.

Diphacinone and chlorophacinone are effective at concentrations as low as 0.005%. Rats may be killed in 3 days and mice in about 6 days.

Non-anticoagulants. The mode of action of these rodenticides is varied: bromethalin kills by disrupting the central nervous system, cholecalciferol kills by overdose of vitamin D_3 and causing excessive calcium in the blood; zinc phosphide kills when the acid in the rodent digestive system reacts with the phosphide to generate toxic phosphine gas. These rodenticides are effective against rodents that are resistant to anticoagulants. There is little or no bait shyness reported for bromethalin and cholecalciferol.

- *Common Use.* The non-anticoagulants in use today are: bromethalin, cholecalciferol, and zinc phosphide. Bromethalin is considered a single-dose rodenticide because only a small amount is lethal, but the rodent may not die until 4 days after eating. Cholecalciferol also functions as a single-dose toxicant, but it can accumulate in the rodent body and act as a multiple-dose rodenticide.

- *Zinc phosphide* is not a commonly used rodenticide. The baits have a strong, pungent garlic-like odor, which attracts rodents, but deters other animals. Rodenticide-grade zinc phosphide usually comes as a black powder containing 75% zinc phosphide and 25% antimony potassium tartrate, an emetic to cause vomiting if the material is ingested by humans or domestic animals.

Rodenticide Resistance. Since 1971, resistance has been detected in rat populations around the world, especially in locations where warfarin has been used for many years. In some populations, resistance to this chemical has provided cross-resistance to other first-generation anticoagulants, including hydroixycoumarins and indaniones.

Resistance to any chemical is a genetic phenomenon that occurs from lengthy, multi-generational exposure to a specific pesticide. The resistance gene exists naturally in populations, but in less than 5% of the individuals. Where rats are controlled primarily with rodenticides, a few individuals survive after eating bait because they have natural resistance, not because they ate a sub-lethal amount. After several generations of constant exposure to the rodent bait, there are more resistant individuals because the bait eliminates susceptible ones. Eventually the population is made up of a large percentage of resistant rats.

CHAPTER 17

INSECTICIDE APPLICATION EQUIPMENT

By Cecil E. Patterson

A technician carrying a stainless-steel tank with a brass valve and extension is the image of professional pest control service most recognized around the world. The expectation is that the combination of a trained individual applying an effective insecticide with specialized equipment will result in long-lasting control.

The worldwide availability of professional application equipment is a major factor in the global standardization of household insecticides. Manufacturers prepare product labels expecting that the service technician will have the training on and access to technically accurate equipment for safe and proper application. Modern insecticides are effective at low concentrations and limited placement, and professional pest control equipment meets these criteria. The majority of equipment for household insecticides is designed and built specifically to match the delivery rates and spray patterns required on product labels.

LIQUID APPLICATION

Most insecticides for use against household and structural pests are applied as a liquid spray. These chemicals are effective when delivered as small droplets to exposed substrates, or as a stream through a crack-and-crevice straw into a narrow void. The most common delivery method is with a compressed-air sprayer that can be carried by the operator and pressurized by hand. Gasoline- or electric-powered equipment may be used for specialized applications, such as power misting for outdoor mosquito control and fogging.

Tank Sprayers. Tank sprayers made of stainless steel or plastic with brass components are commonly used for residential and commercial pest control. They are available in 0.5-, 1-, 2-, and 3-gallon sizes; their capacity, weight, hand-pressurizing, and specialized nozzle make them suitable for most treatments.

- *How they work.* The tank acts as a pressurized air chamber. The pressurized air forces the liquid from the tank to the valve, then to the nozzle. The tank is filled to about 3/4 capacity to allow sufficient head space to pressurize a volume of air. A plunger-type pump is used to introduce and compress air in the space above the liquid. The tank usually has a siphon tube to move the liquid to the nozzle through a hose. A chemical-resistant hose carries the liquid to the nozzle through an extension lance or wand. The extension may be 8 to 24 inches long, with the end straight or angled at about 30°. The tank may be fitted with a pressure gauge, which ensures proper air pressure is maintained for efficient use of the nozzle.

  During use, it is necessary to periodically pump air into the tank to maintain operating pressure. Pressure in the tank decreases as liquid is discharged and the volume of air in the tank increases. This decreases the output and spray angle of the nozzle and increases the droplet size.

- *Selection.* Tank sprayers are available in 0.5-, 1-, 2-, and 3-gallon sizes; these are practical and can be hand carried. Selection is typically based on the intended use, such as small tanks for indoor application and large tanks for outdoor treatments. The most ergonomic is the 1 gallon. The mixing instructions and dose-packaged formulations for most household insecticides are prepared for a 1-gallon final dilution.

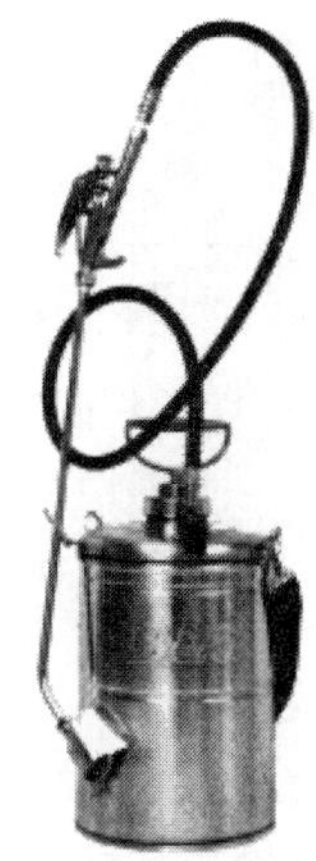

Compressed-air sprayer

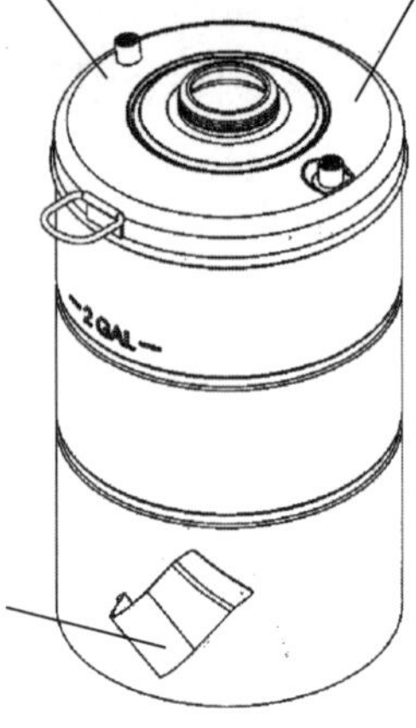

Compressed-air sprayer

Compressed-air tanks are actually larger than their operational volume; this permits head space at the top of the tank for compressed air. Fill marks embossed on the side of the tank indicate maximum capacity. Filling above this mark reduces head space and requires the operator to pump frequently to re-pressurize the tank. The initial head space needed for a 1-gallon tank is 3 pints; the initial head space for a 3-gallon tank is about 8 pints.

The fill lines are marked on the side of each tank. The embossed circular ridges in the tank are not fill lines; they are for strengthening the tank.

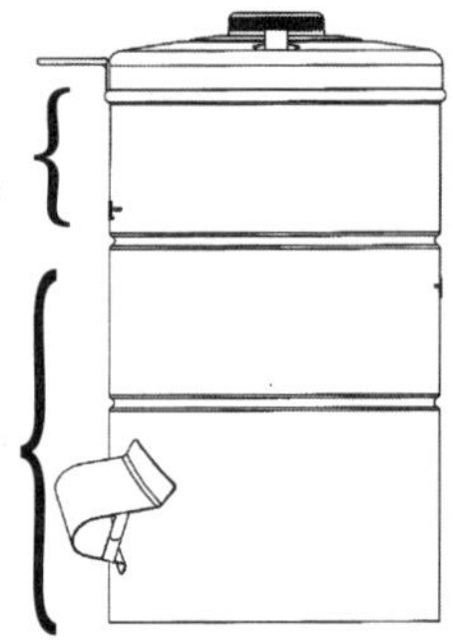

Tank selection can be based on safety and performance features, such as valve, length of the extension, carry strap, nozzle, pressure gauge, or pressure-release valve. Some of these may be required by regulatory agencies or company policy; others may be for operator convenience.

Bottle Sprayers. Small versions of compressed-air sprayers have a capacity of 1 quart or less, and the tanks are metal or plastic. Most models operate on the same principle as large sprayers, having a pump cylinder and siphon tube, with the valve incorporated into the handle. There is limited head space in the tanks, and frequent pumping is necessary to re-pressurize. The nozzle may be at the valve or at the end of a short extension.

Bottle sprayer

- *How they work.* Bottle sprayers have the same mechanics as standard spray tanks. The pump mechanism includes a plunger cup and check valve, and the mechanism is repairable on most models. About 10 pump strokes provide operational pressure, and the sprayer can be quickly re-pressurized with just a few plunger strokes.

- *Selection.* Technicians typically carry small bottle sprayers in a belt pouch, so they are within reach when needed. The benefits of this size include light weight, the ability to prepare only a small amount of a finished insecticide, and the ability to change pesticide by changing to a different bottle. Additional bottles can be prepared, labeled, and quickly installed. These sprayers are suited for limited applications, for use in a confined space where a standard tank would be inconvenient, or for use in a location where a small, discrete sprayer is preferred.

These small sprayers are typically used for spot treatments, because application rates can be much higher than with standard tank sprayers.

Hand Sprayers. Hand sprayers are plastic, 1-quart, hand-carried sprayers. They have a simple pump mechanism and typically a cone-spray nozzle. The nozzle may be adaptable for fan-spray, crack-and-crevice, or foam delivery. The combination handle and trigger make them practical for treating indoor and outdoor surfaces. They are durable and inexpensive, but their daily use is limited. These small sprayers are often used for specialty liquid applications, such as foam treating of floor drains with a biological cleaner, or application of single-purpose insecticides for bed bug or cockroach control. For some insecticides, chemically resistant gaskets are required.

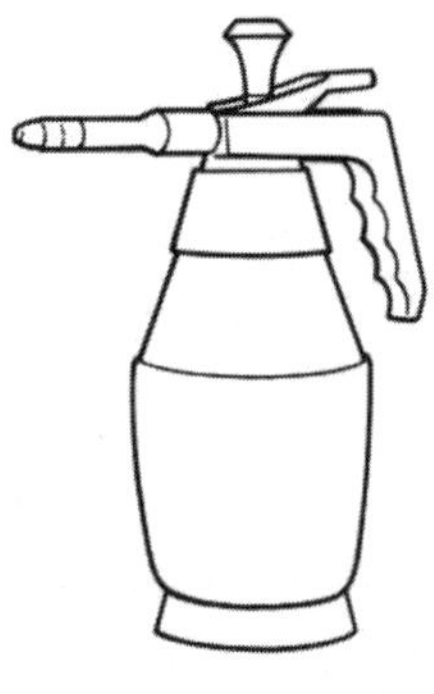

Backpack Sprayers. These are large metal or polypropylene-plastic tank sprayers. They are used primarily for treatment of vegetation and application to building perimeters. Knapsack or backpack sprayers are usually shaped and fitted with straps, so they can be carried on the operator's back. The operator can simultaneously operate the side-mounted pump handle with one hand and direct the spray with the other. The hose, valve, and nozzles are the same as those used on compressed-air sprayers, except that models for agricultural use have cone-spray nozzles. The pump is either a piston or diaphragm mechanism; battery-pressurized models are also available.

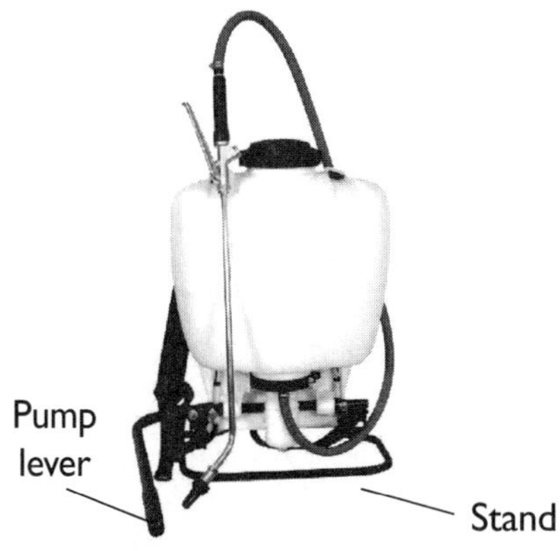

Backpack sprayer

- *How they work.* Delivery is through a pump cylinder activated by the external pump lever. On the up-stroke, liquid from inside the tank enters the cylinder, and on the down-stroke this liquid is forced through the hose and valve. Diaphragm pumps operate by pressure created by a flexible diaphragm compressing liquid in a chamber and from there to the hose and valve.

- *Selection.* The 3- or 4-gallon capacity of backpack sprayers means the operator must carry 24 to 32 pounds while walking and spraying. Sprayers with piston pumps are normally used when high pressures at the nozzle are required. Sprayers with diaphragm pumps are used when insecticide formulations may damage the piston chamber. Some piston-operated backpack sprayers are equipped with coarse and fine fan nozzles that meet insecticide label requirements for perimeter application. The pump mechanism is relatively smooth and resistant to wear, especially with use of wettable powder formulations.

Cone-spray nozzles are commonly found on backpack sprayers, but they are not appropriate for surface applications. A cone spray has a circular pattern, which is suitable for treating plants but not for the horizontal and vertical surfaces needed for household and structural insecticide application.

Backpack Mist Blowers. These sprayers are used for treating large outdoor areas, specifically for penetrating ground vegetation and tree foliage with insecticide. The force of the air stream and the range of droplet sizes provide effective control of flying and resting insects and can create a droplet residue on surfaces. The most common use is for control of mosquitoes.

- *How they work.* The motorized mist blower typically uses a two-cycle motor and an engine-driven fan to produce a high-velocity air stream that delivers liquid insecticide. The air output of the fan is directed to the nozzle through a delivery hose. The tank is pressurized to force liquid to the nozzle. The nozzle is mounted at the end of the delivery hose so that the droplets produced are entrained in the air stream. Spray droplets are projected about 30 feet horizontally and 20 feet vertically. The droplet size range for mist blowers is 51 to 100 microns. These sprayers are used for outdoor application, and some models allow delivery of liquid, powdered dust, and granular material.

Mist blower

Electrostatic Sprayers. Electrostatic sprayers have been adapted for indoor use to apply insecticides and disinfectants. These sprayers apply electrically charged droplets to improve surface delivery and spray coverage. However, indoor electrostatic spraying does not always provide increased deposition or coverage compared with that of conventional aerosol sprays. There is limited penetration of charged droplets into the multi-level substrates that are common indoors. Deposition is best on surfaces close to the nozzle.

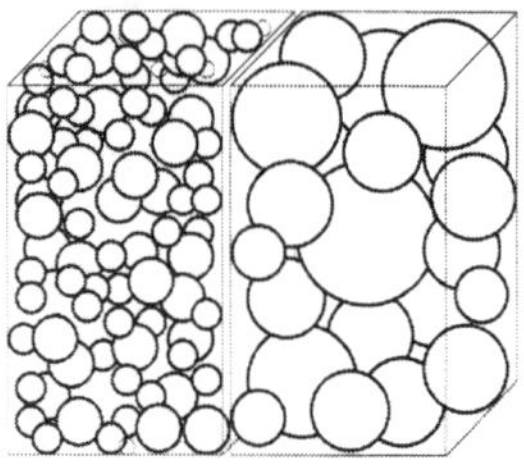

In volume median diameter (VMD), half the droplets in the spray are larger and half are smaller than the VMD.

- *How they work.* The spray is driven at high velocity; as droplets exit the nozzle they pass through a charged electric field and attain a negative (-) charge. At close distances, positive (+) grounded substrates attract these negatively charged droplets. Once droplets contact a surface, they lose their charge. Droplets from electrostatic sprayers have a range of sizes, but nozzles are typically designed to deliver 40 microns (VMD).

- *Selection.* Electrostatic delivery may have limited use for household pest control applications. For electrostatic attraction to take place the target surface must be grounded and have a positive charge to attract the negatively charged spray droplets. If not, the droplets will act much like conventional spray droplets in their deposition and coverage. Few indoor surfaces have or maintain a charge suitable for electrostatic sprayers to be effective.

Protective coveralls, eyewear, and respirator are recommended when electrostatic sprayers are used. The valve and extension on these sprayers is short and the force of the air-assisted spray and the charged droplets can result in applicator exposure.

The VMD droplet size listed for foggers does not represent the average size droplet produced. Only about 10% of the droplets are this size, 50% of the remaining droplets are larger, and 50% are smaller than the VMD number. Fogger nozzles can not deliver one droplet size, there is always a range in the droplets produced.

FOGGING

Fogging is generally used to describe the application of an aerosol spray with droplets in the range of 10 to 20 microns (VMD). Manufacturers may describe foggers as delivering a certain droplet size, and foggers are often selected on this basis.

Cold Foggers. Indoor space spraying is usually done with small, 0.5 to 1.5 gallon, cold foggers. The droplets produced are formed outside the nozzle and carried in an air stream about 3 to 6 feet from the nozzle. The droplets produced are in the range of 20 to 56 microns (VMD), depending on nozzle design and flow rate. As with any fogger nozzle, there is a range in droplet sizes produced, including a small percentage that are less than 20 microns.

- *How they work.* These machines produce droplets mechanically by shearing or breaking apart a flow of liquid insecticide with a strong air current. They usually have an electric fan which forces air through the nozzle. Liquid insecticide in the tank is fed through a nozzle where it is broken up into small droplets. The air velocity remains constant, but the flow of liquid to the nozzle may be adjustable. Decreasing flow produces small droplets, increasing flow produces large droplets.

- *Selection.* There are two basic types: those with a stationary tip, and extendable-tip models in which the tip is at the end of a flexible hose. Stationary-tip models have the tank and spray nozzle combined into one unit. In use, they are placed centrally in the treatment space and remain stationary during treatment. Some have the capacity to adjust droplet size, while others are for basic treatments and are not adjustable. A stationary-tip fogger is suitable for open spaces where the fogger can remain unattended during operation. Extendable-tip foggers have an application hose for in-hand use so that the nozzle can be directed to specific locations.

Cold fogger

Thermal Foggers. Thermal foggers are generally used outdoors for mosquito control. They are not used to treat enclosed or non-ventilated spaces because of the potential for an explosion resulting from exposure to a flame, spark, or electrical appliance. Thermal fogging has been used to treat for cockroaches in sewers; in these cases the fog is directed into a manhole from above.

Proper protective equipment must be worn when using thermal foggers; this may include a full-face respirator for some insecticides.

Large, pulse-jet foggers typically have a long, 70-inch profile and an operating weight of about 75 pounds. The pressurized formulation tank usually holds 3 gallons; this capacity permits extended application time. The formulation output is adjustable, usually to a maximum of about 18 gallons per hour; the droplet size can be adjusted, but typically, it is about 50 microns.

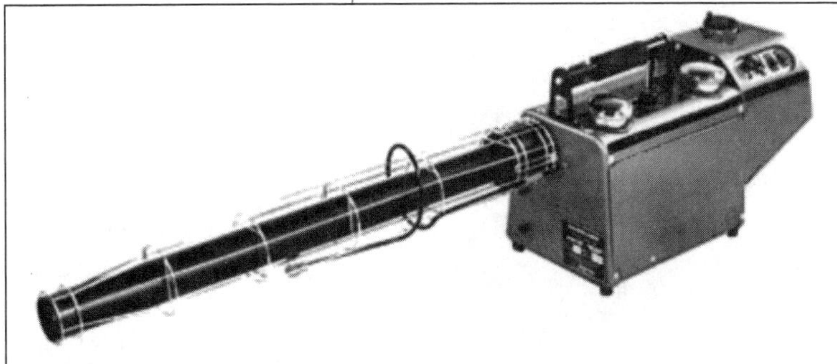

Thermal fogger

- *How they work.* A thermal fogger produces a flow of hot air into which a dilution of insecticide is delivered. The velocity of the air shears the liquid oil into droplets and a temperature of more than 500°C vaporizes the droplets. When the hot gas passes through the extended exhaust pipe into the air, it condenses to form a dense white fog.

  In pulse-jet foggers, an initial mixture of fuel and air is delivered to a combustion chamber and ignited. The gas escapes the combustion chamber at high velocity through a long pipe of a smaller diameter than the combustion chamber. When operating with the proper mixture of fuel and air there are about 80 pulses per second. Liquid insecticide dilution is delivered through a valve to a nozzle, or "jet," in the exhaust pipe. For units using oil-based dilutions, the inlet is close to the combustion chamber to give more complete vaporization of the liquid.

AEROSOL APPLICATION

Applying an insecticide as an aerosol reduces the amount of active ingredient used, while delivering it directly to infested harborages. Crack-and-crevice application is the most common and effective method of aerosol use. Although manufacturers provide an actuator or spray tip that accepts a crack-and-crevice straw, most professionals use an applicator that extends the original spray tip and hand-grip on a coiled hose.

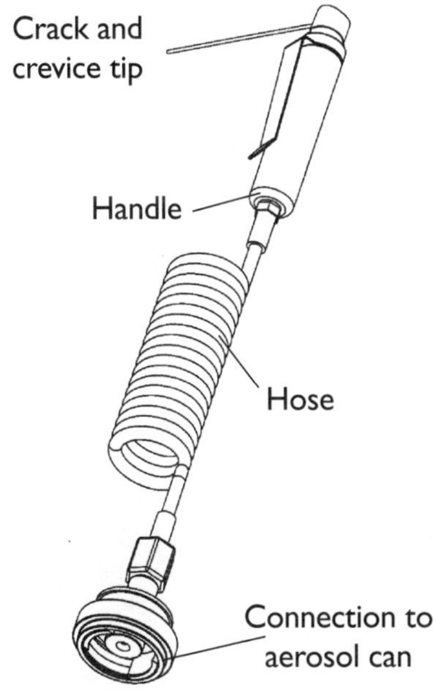

Aerosol Delivery Systems. Delivery systems designed for aerosol cans extend the spray tip from the can with a coiled hose. This improves the ability of the applicator to spray confined surfaces and treat cracks and crevices that may be inaccessible to conventional sprays. Aerosol delivery systems typically increase the efficiency of an aerosol can by decreasing the amount of liquid delivered per minute.

- *How they work.* A plastic or metal clamp fits on the aerosol can top and engages the valve stem. A ball valve above the clamp opens and closes the liquid flow from the can into the hose connected to the handle. A standard aerosol spray tip or the tip from the aerosol can is used for treatment. This maintains the flow rate and spray delivery pattern from the original can. These delivery systems include a belt pouch for carrying the unit.

- *Selection.* The selection of an aerosol delivery system is usually based on convenience for the operator and compatibility with the major aerosol can manufacturers. Attaching the delivery unit to a new aerosol can is often necessary during service, and it must be done quickly and efficiently. The weight and grip of the delivery-system handle are important considerations. The dimensions of the valve stem at the top of the aerosol can depend on the manufacturer.

Portable aerosol sprayer, cart model

Portable Aerosol Sprayers. Portable aerosol sprayers have a directed air stream with aerosol droplets to provide the opportunity to treat small and large spaces for flying- and crawling-insect control. Systems are designed to be adjustable to deliver a range of droplet sizes for use in crack-and-crevice treatments or to apply its air-only feature for flushing pests without chemicals.

- *How it works.* The aerosol droplets are produced inside the nozzle and exit in a strong air stream. The droplet spray can extend 4 to 6 feet from the nozzle. A long, coiled hose and 9- to 18-inch extension on a valve make these sprayers suitable for treating large spaces and narrow voids. The tank capacity for these sprayers ranges from 28 ounces to 1 gallon; cart-mounted units typically have 3-gallon tanks. A pressurized tank provides continuous or adjustable flow of liquid and air to the valve. Droplet size can be adjusted by regulating the flow of liquid to the valve: reducing flow decreases droplet size, increasing flow increases droplet size of a power mist. Air pressure in the tank is pre-set and maintained by an on-demand compressor.

- *Selection.* Preferred features include portability and ability to alternate between crack-and-crevice application and aerosol delivery. Emphasis on small droplet size is unnecessary, since the majority of use is for general pest control and not space spraying/fogging. Durability and maintenance are important considerations as these sprayers will become a standard application tool.

DUST AND GRANULE APPLICATION

Dust and granule formulations are applied with hand-operated units that permit some regulation of the amount delivered. There is limited calibration for these tools, because hand use is irregular, and the particle size of dusts and granules is variable. Compressed-air dusters deliver large quantities of dust. Motorized, battery-, or electric-powered dusters may dispense metered amounts, and hand-held piston-type dusters can deliver consistent amounts. Powered granular applicators typically have a large reservoir and operate by entraining granules in an air stream for delivery.

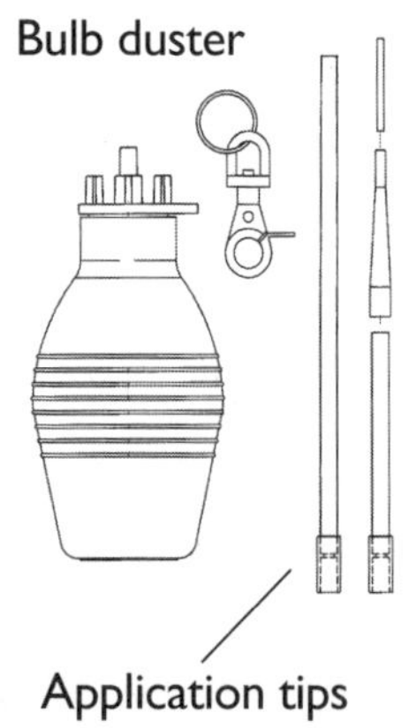

Aerosol dust formulations are available for a limited number of active ingredients, including boric acid and silica. The aerosol-can delivery system moves the dust through plastic tubes which may impart an electrostatic charge to the particles.

Dust Applicators. The objective of dust treatment is to apply a thin film to unexposed sites. In general, modern dusts resist moisture and clumping and flow easily from equipment chambers through application tubes.

- *How they work.* A pulse of air is directed into a chamber to agitate dust particles into the space, or the chamber is agitated by hand. The result is that air is expelled from the chamber carrying fine dust particles. Hand dusters are positioned so that the discharge tube is expelling the air-and-dust mixture from the top of the chamber. If the duster is positioned so that the discharge tube is immersed in the material, a large amount of dust will be delivered.

  In bulb and piston dusters, the pulse determines the force of the air leaving the discharge opening and the distance the dust is spread. The piston stroke or the squeeze of the bulb limits the dispersal of dust to 10 to 12 inches.

  In turbulent-air models, the dust is in a large closed chamber, and air is forced into the chamber. Pressure and turbulence forces air carrying dust through a discharge hose to a tip or nozzle. The velocity of air leaving the chamber through the hose is sufficient to disperse dust 6 to 10 feet from the end of the tip. Battery-operated hand dusters work on the same principle, except the chamber is small and the air velocity is less. Compressed-air models are based on a pressurized container, such as a stainless-steel tank or metal bottle. Air carrying dust is discharged under pressure. Some agitation of the dust in the system is required.

The use of some silica-based products can permanently damage plunger-type dusters. The fine particles can penetrate the pump cylinder and damage the pump gasket.

- *Selection.* Selection and use of a hand-operated or motorized duster is based on its ability to treat horizontal and vertical surfaces with a continuous layer of dust. Application tips facilitate application into voids, crevices, and galleries of termites, wood-infesting ants, and carpenter bees. Occasionally, large enclosed areas, such as attics and crawlspaces, must be treated, and these require a power duster capable of dispersing a large amount of dust.

  - **Bulb dusters** benefit from a pliable dust chamber for smooth and precise delivery. Models made of natural rubber retain flexibility and resist insecticide and UV-light damage. Plastic crack-and-crevice application tips provide safety when treating electrical outlet and fuse boxes. The typical insecticide-dust application rate with the B&G Model 1150 bulb duster is 0.04 ounces per squeeze.
  - **Bellows dusters** are small and typically held upright when used. This positions the discharge tube at the bottom, which can result in applying a large amount of dust.
  - **Piston dusters** typically carry more dust than bulb dusters and are readily calibrated because of the consistent pump stroke. The swath for the B&G Model 1151 piston duster is about 38 inches long and 8 inches wide. The large piston duster (B&G Model 1152) has a chamber that holds 1.5 pounds of dust; the application rate is about 0.25 ounces per stroke.
  - **Electric dusters** hold several pounds of dust and are used primarily to treat large spaces, such as attics and crawlspaces. The swath for the B&G Model 2250 electric duster is about 20 feet long and 5 feet wide for coverage of about 100 square feet.

Piston-type duster

- o **Battery-powered dusters** are trigger operated and have adjustable flow (grams per second) and are best suited for crack-and-crevice application.

Granule Applicators. The most common granular application is outdoors for perimeter control of insects. Granular baits will be gathered and moved by the target pest; contact-control baits will remain in place and interact with the pest or the soil. Granular formulations differ widely in size and shape. The weight of insecticide granules is variable, and this influences delivery.

Product labels also vary in terms of application rates: they may be in ounces or grams and specify delivery per square yard, per 100 square feet, or per linear foot. Calibrating the applicator for each product is essential for label compliance and to prevent costly over application.

- *How they work.* Most dust-application equipment can be used to apply granules. However, because of the particle size of some granule formulations, the use of bulb and hand dusters is limited. When applying granules with a hand duster the siphon tube in the chamber must be immersed in the granules. Increasing pressure in the chamber, by squeezing a bulb duster or pumping a plunger duster, forces granules out the delivery tube.

  Electric-powered granule applicators usually operate by forcing air into the chamber, and the turbulence moves some granules into the air stream, and pressure in the chamber forces the air-and-granule mix through the delivery tube(s).

- *Selection.* The application objective for granules is uniform coverage of the surface. Selecting a granule applicator may be based on its ability to spread a specific product outdoors or to treat entire attics or crawlspaces. Although most dusters can also be used to apply granules, it is best to dedicate pieces of equipment for one or the other. Contaminating granular bait with an insecticide dust or insecticide granule can be counterproductive and costly.

  - o **Hand dusters** can be used to apply most small-particle granule formulations. These granules flow easily through the siphon and application tube, but not a crack-and-crevice tip. The treatment swath for a typical insecticide granule formulation of small particles using the B&G Model 1150 bulb duster is about 5 feet long and 1.5 feet wide, and the treatment zone is about 7.5 square feet.
  - o **Piston dusters** are preferred for delivering granules outdoors because the application tip ensures efficient placement. The treatment zone for the B&G Model 1151 small piston duster delivering a granular formulation is about 7.5 square feet, the swath is 18 inches wide and extends about 5 feet from the application tip. The delivery rate is variable and depends on the product; with a small-size granule, the rate is 0.06 ounces per stroke of the plunger.
  - o **Electric granular applicators** hold a large amount of product and are suited for treating large spaces, because they have a 22-mph air velocity that can project even large granules 10 feet or more. The swath for applying granules from the B&G Model 2250 portable unit, fitted with an extension tube, is about 15 feet long and 36 inches wide; the coverage area is about 50 square feet.

GEL BAIT APPLICATION

Gel formulations of baits for cockroaches, ants, and termites have nearly the same consistency. The bait matrix is gelatinous or a semi-solid material that has high water content, but retains its shape after application. Gel baits are usually applied from plastic syringes, either directly or by means of a pistol-type tool.

- *How they work.* Pistol-shaped gun applicators with a finger-trigger mechanism are used to disperse gel bait from plastic syringes. Bait syringes have a common design and can be attached to all commercial bait guns. Most guns operate by the gradual advancement of a plunger rod into the back of a bait syringe. Pressure from the rod forces the bait out the tip of the syringe; the amount of bait depends on the movement of the plunger rod. For dose-specific guns, the movement is controlled by pre-determined settings, and the amount is approximately the same with each trigger pull. For variable-dose bait guns the amount of bait delivered is controlled by the operator and each trigger pull. This variability can result in costly over application.

  Vacuum-pressure guns provide consistent output of bait, with adjustable deposit amounts with each trigger pull. The design includes check valves to ensure tip shut-off.

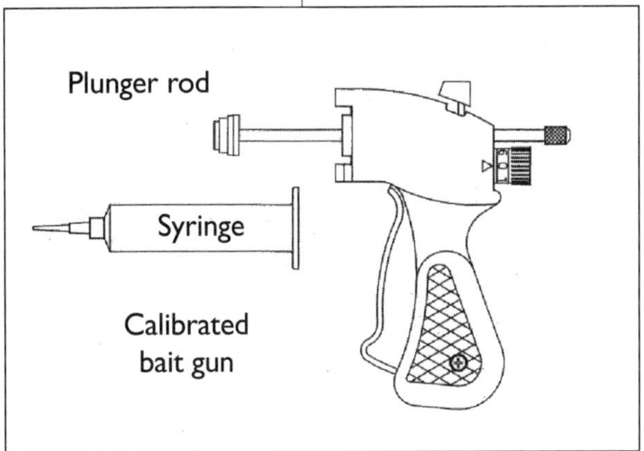

  o *Selection.* Bait product labels recommend a specific amount or size of placement for control and cost effectiveness. Dose-specific bait guns can be adjusted to deliver the same amount of gel with each pull of the trigger; the amount of bait can not be changed by partially pulling the trigger. The adjustable settings provide a range of placement sizes and amounts to comply with label recommendations. These bait guns ensure consistent delivery, and with that they maximize service time and control costs by ensuring the efficient use of each bait syringe.

FOAM APPLICATION

Foam application is a way of delivering a small amount of liquid to a surface or confined space, such as a wall void. Termiticide foam can flow over the uneven soil surface in the void beneath concrete slabs. It provides uniform coverage of the soil surface, and it can flow around obstacles and utility service pipes. Termite galleries in structural wood can be treated by drilling and injecting below the wood surface; foam can spread through the open galleries. The low expansion ratio of wet foam provides a short application time and quick collapse of the foam to the soil.

- *How they work.* Foam-application equipment converts water dilutions of insecticide and a surfactant into foam by introducing air into this mixture before it exits the nozzle. Variations in foam-application equipment are usually in the means or mechanism of introducing air into the liquid insecticide. Typically, air is introduced before the liquid enters the application hose. In small hand sprayers that have a foaming tip, the air enters as the liquid leaves the nozzle. Variability in the produced foam may be due to the amount of surfactant in the mixture, the solvents in the insecticide formulation, and the amount of air introduced into the liquid stream.

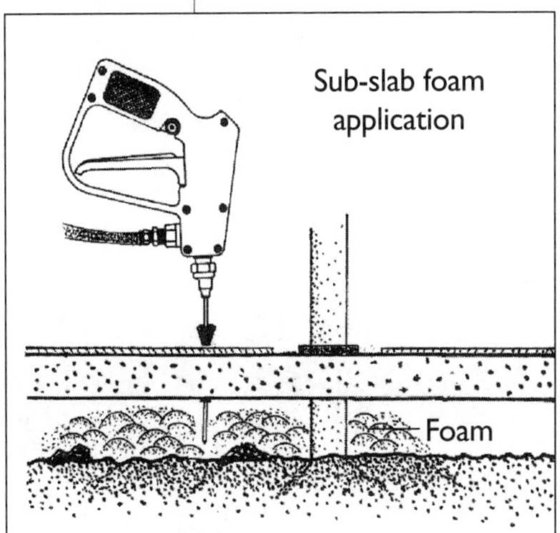

- *Selection.* Foamers are available in different configurations from a power foamer, which uses an electric pump to pressurize the tank, to a portable hand foamer. The power foamer uses an air compressor to supply the air required to make the foam. This size foamer is suitable for treatment of large surfaces and wall voids; sub-slab application of termiticide foam; and treatment of other locations in which a continuous stream of foam is needed. The hand foamer is based on a 1-gallon, stainless-steel tank and uses the compressed air created by pumping the sprayer to create foam. This unit requires regular pumping for repressurization. The hand foamer is suitable for spot treatments, such as floor drains, and galleries of wood-infesting insects.

  - **Wood and other porous surfaces** can be effectively treated with insecticide foam because penetration is enhanced by increasing the amount of liquid (as foam) retained on the surface. The objective of treating exposed and enclosed wood surfaces is to establish a residue when the liquid insecticide drains from the air-bubble matrix.
  - **Biological cleaners** can be applied as foam to grease-covered surfaces and to drains clogged with organic debris. For horizontal and inverted surfaces, foam spreads and remains on the surface to increase delivery of microbes in the dilution; on organic matter, foam increases the liquid surface area for better distribution of microbes on the substrate.

TERMITICIDE APPLICATION

Below-ground application of liquid termiticides is through directional nozzles attached to a soil pipe and a low pressure pump of about 25 psi. Modern termiticides require continuous coverage in the treatment zone. This is achieved by uniform spacing of the soil pipe to ensure overlap of the spray pattern from a 180° or 360° nozzle. For sub-slab application, pipes are inserted through holes drilled in the surface; treatment is made with a directional or straight tip to the soil surface or below. Specialized tools permit sealing of the hole in the concrete around the treatment pipe to eliminate splash-back of liquid to the surface above the hole.

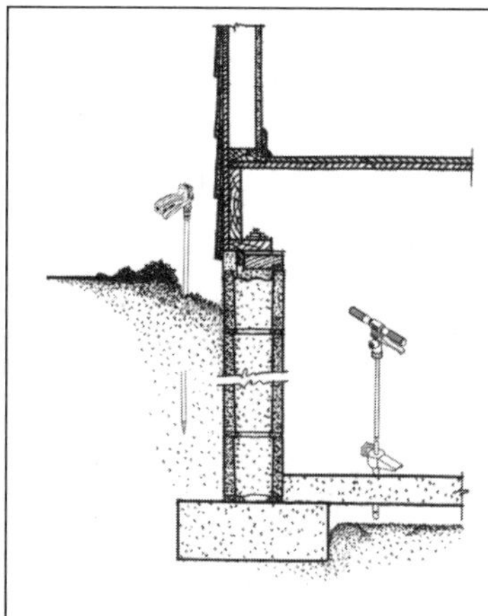

Termiticide application

- **How they work.** The combination of valve, soil pipe, and tip are designed to deliver insecticide to the soil adjacent to foundation walls. Trigger- or lever-action valves regulate the flow of liquid from the mixing tank. At the end of the soil pipe is a 180° fan, 360°-four-way, or straight tip; the flow rate for each tip is 2 gallons per minute. The lateral spread of the spray pattern for the fan and four-way tip is about 12 inches on either side of the nozzle and about 3 inches below the tip; the spread for a straight tip is about 6 inches.

  Sub-slab injection tools are designed to seal the opening around the treatment pipe in the concrete. This is done with a rubber gasket which expands in the hole and around the pipe after it is in place. The gasket serves to prevent the splash-back of liquid termiticide onto concrete floors covered with tile, wood, or other material. Sub-slab treatment rods are equipped with an adjustable foot ram that includes a rubber seal that closes the top of the hole drilled in the concrete slab to prevent splash-back from termiticide applied beneath the slab.

- *Selection.* The professional termiticide application tools that establish a continuous treatment zone are those with a reliable valve and durable soil pipe(s), and a calibrated directional nozzle. This set allows the operator to accurately space insertions of the soil pipe to achieve a continuous band of treated soil. Performance of modern termiticides is linked to proper application with professional tools and trained operators.

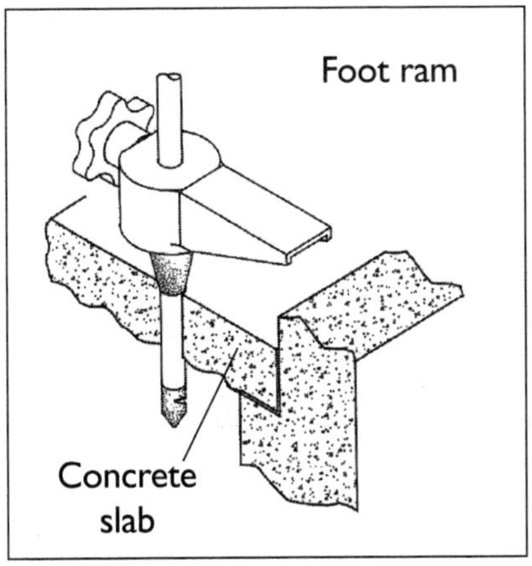

CHAPTER 18

EQUIPMENT MAINTENANCE AND REPAIR by Claude Thomas

Although modern application equipment is made of durable materials, normal use results in wear and a gradual decline in performance. Without routine inspection, maintenance, and replacement of key parts, any delivery system will become inconsistent and waste insecticide. Broken and leaking equipment is not only non-profitable, it can result in misapplication, contamination, and insecticide exposure of non-target organisms. All of which may result in regulatory action.

Routine maintenance is primarily the responsibility of the individual applicator. Changes in performance will be detected first by the applicator and are best repaired by that individual. A *prevention-based* maintenance program will eliminate breakdowns and repair while on the job. A *failure-based* program is costly and dangerous to applicators and customers.

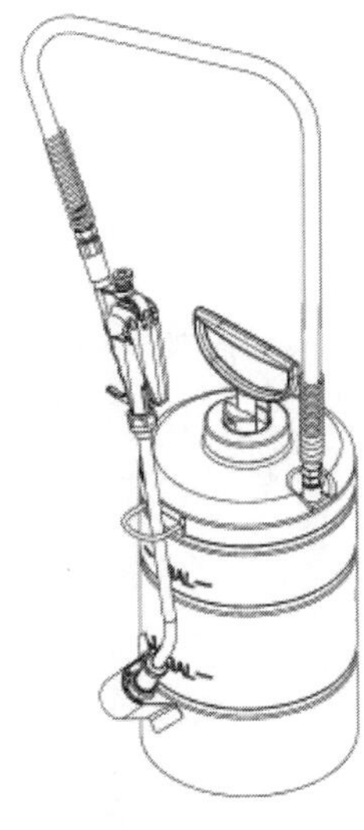

Compressed-air sprayer

COMPONENTS AND DURABILITY

Pesticide-application equipment is made from durable components, including stainless steel, brass, molded polyethylene, and polyacetal plastic. The seals and gaskets are synthetic rubber or other elastic material. All of the components are specifically designed to be compatible with the solvent systems of modern insecticides and formulations. However, equipment manufacturers cannot anticipate all the solvents that may be used in spray equipment.

Seals and Gaskets. Flexible seals are positioned at metal-to-metal connections to prevent leaks when liquid or dust is under pressure. The seals used in application equipment are made of nylon or synthetic rubber. The types of man-made rubber include, ethylene propoylene diene M-class rubber (EPDM), Viton™, and Teflon™.

- *Nylon.* This material has high tensile strength, elasticity, and a low compaction rate. It resists abrasion and tolerates most organic fuels, oils, and chemicals. However, nylon is not generally compatible with strong acids.

- *Viton.* This copolymer material is compatible with most solvents, but is not resistant to acetone and acetic acid. Viton Extreme is compatible with acetone and acetone-based materials.

- *Teflon (PTFE).* This material is non-stick, and water and oil resistant. The coefficient of friction is the second lowest of any solid material, and it is used where sliding action occurs. PTFE has no "memory" and is subject to compression set—that is, once it is compressed it may retain that shape. However, a small amount of compression allows PTFE to conform to surfaces and form a tight seal.

- *Synthetic rubber.* Buna-N rubber (Nitril rubber) is more chemical resistant and durable than natural rubber. Buna-N is resistant to oil and most chemicals, but it does not resist acetone, toluene, xylene, or benzene.

- *EPDM.* Seals made of EPDM rubber are used in electric-pump motors of

Silicone rubber gaskets resist most chemicals, but they have a very low tensile strength, and their replacement interval is short.

Borate formulations using ethylene glycol or other alcohol as a solvent will damage EPDM seals in sprayers.

application equipment and sprayers. EPDM is compatible with hydraulic fluids, acetone, and strong alkalis. It is not compatible with concentrated acids, gasoline, kerosene, xylene, and ethylene.

Oil-based insecticides for spraying or fogging often use deodorized kerosene as the carrier. These formulations will damage EPDM seals in the equipment.

Tanks. Tanks are made of stainless steel or polyethylene.

- *Stainless steel* has a minimum of 10% chromium. This component forms a thin layer of chromium oxide on the surface. The oxide layer is impervious to liquid and air and protects the metal from corrosion. The oxide layer quickly re-forms after the surface is scratched to ensure protection.

Stainless steel is susceptible to chloride pitting and can be damaged by cleaning liquids that contain chlorine bleach.

Modern tanks have creases in the steel to provide strength and stability. The base of the tank is recessed to prevent contact with the surface, and a concave bottom provides compression stability. Molded-plastic tanks are seamless to eliminate the possibility of leaks. Stainless and plastic tanks can withstand pressures well above that obtained by hand pumping.

- *Plastic.* Polyethylene is widely used for tank sprayers, backpack sprayers, and small hand sprayers. The chemical resistance of commercial-grade polyethylene includes the acids: formic, nitric, and sulfuric; and alkalines: potassium hydroxide and sodium hydroxide. It is not compatible with acetone, benzene, carbon tetrachloride, ethyl acetate, methylene chloride, trichloroethylene, and turpentine.

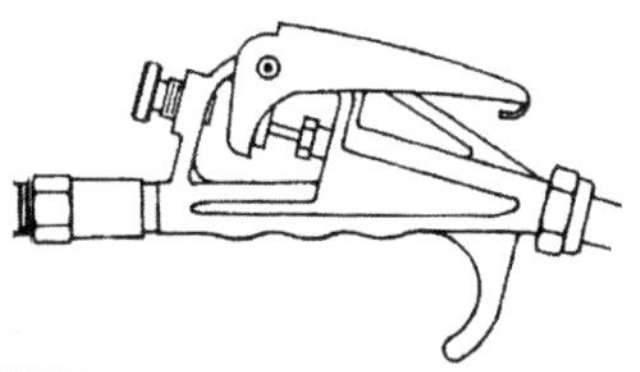

Shut-off valve

Valves, Pump Cylinders, Siphon Tubes. Shut-off valves on tank sprayers are forged from brass and stainless steel, but some are made of polyethylene. In most sprayers, the valve is under constant liquid pressure once the tank is pressurized. When the shut-off valve is closed, it prevents the flow of liquid to the nozzle until the trigger is depressed.

Pump cylinders and siphon tubes for professional tank sprayers are made of seamless brass tubing to prevent cracks and provide chemical resistance. The handle, pump rods, and plungers are made of brass, stainless steel, or polyethylene plastic.

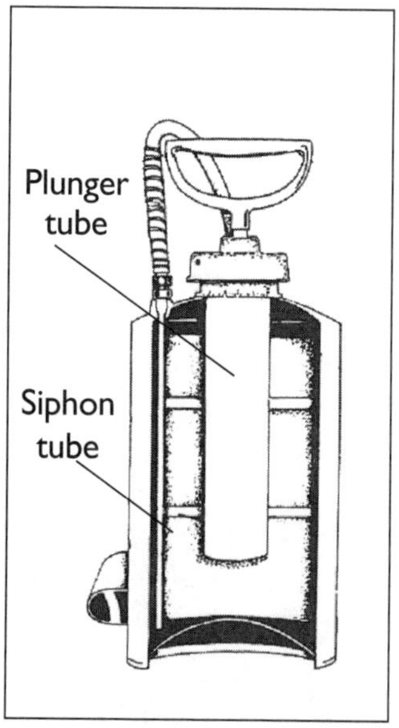

Nozzles. Brass nozzles are used in professional pest control. This material is economical, but it is subject to wear and corrosion from ammonia-based liquids and disinfectants with 7% or higher concentrations of hydrogen peroxide. Brass is abraded by dirt particles in the spray liquid and by wettable powder formulations.

Filters and Check Valves

- *Filters.* Sprayers and foggers have one or more stainless-steel screens or filters along the path of the liquid from the tank to the nozzle. These 50-mesh filters prevent soil and other material from reaching the nozzle, where they can block or abrade the orifice.

  A 25-mesh coarse filter may be used in some equipment. It is located at the hose-valve junction or at the end of the extension before the nozzle; typically it is a part of a check valve for an anti-drip tip.

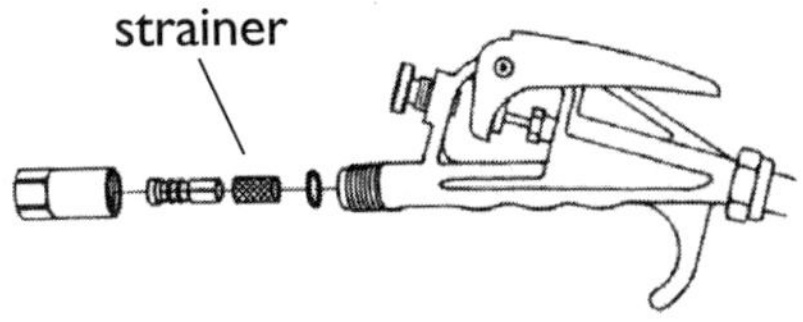

- *Nozzle check valves.* A combination of check valve and filter may be set behind the nozzle. The valve is usually a stainless-steel ball atop a spring, and the unit is surrounded by a filtering mesh. Liquid pressure in the extension pushes the ball to open the flow to the nozzle. When liquid pressure in the extension decreases, the spring in the check valve pushes the ball up to stop flow. Performance is affected by insecticide residue on the steel ball and spring. Cleaning the mesh filter and replacing old valves will ensure proper flow to the nozzle.

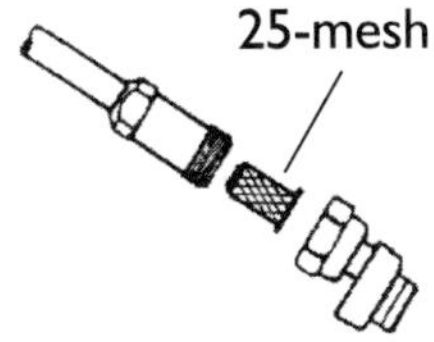

Check valve and filter

The springs in a check valve have a pressure rating for operation. The tank operating pressure needed to open the valve may be 10, 15, or 20 psi.

Hoses. Quality hoses are double-ply and include woven-fiber reinforcement. Typical pressure ratings are 400 psi, well above the operating pressure of the tank. Exposure to extreme heat or sharp edges can damage the outer portion of double-layer hoses and threaten safe operation.

Hoses are made of nylon, polyurethane, EPDM, and polyvinyl chloride (PVC). Nylon is compatible with most acids, alkalines, and solvents. PVC is compatible with most acids, alkalines, glycerine, and glycol; it is not compatible with acetone, benzene, carbon tetrachloride, toluene, or xylene. EPDM is generally not compatible with concentrated acids or solvents.

Generic insecticides may have xylene and other solvents that damage hoses made of polyurethane and PVC.

Pressure Gauges and Regulators. Air pressure in the tank or liquid pressure at the valve can be monitored with a pressure gauge. The components of gauges are sensitive to condensation and vibration, and their service life can be limited. Liquid-filled gauges have few moving parts and long service lives.

- *Pressure-release valves.* These safety valves regulate the maximum pressure that can be achieved in a pressurized container. They are pre-set to a specific pressure, which is usually indicated on the valve. When that pressure is achieved, the valve opens to exhaust the container; it then closes immediately and is available for the next pressurizing cycle.

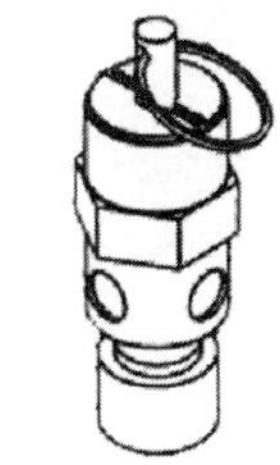

Pressure-release valve

Liquid from the tank may enter the valve during pressurization and opening and result in faulty operation of the valve.

NORMAL WEAR UNDER SERVICE CONDITIONS

All mechanical tools have an expected pattern of wear with normal use, and pesticide application equipment is no exception. Although stainless steel, brass, and seals are very durable, maintenance and replacement is expected with normal use. Equipment manufacturers provide guidelines, diagrams, and parts kits to encourage routine maintenance. Use of original parts ensures that equipment functions properly and maintains the manufacturer's warranty.

Damage to the top screw threads, the port for the siphon tube and hose, or the port for the pressure gauge or relief valve necessitates tank replacement.

Tanks. Residues from insecticides and foaming agents build up in hoses, valves, and extensions, but especially in tanks. When residues accumulate on internal surfaces, equipment performance decreases. Regular flushing will limit the buildup of pesticide residue, but thorough cleaning is periodically necessary.

When dents occur along the welded seam at the side of the tank, the strength and integrity of the tank under pressure may be reduced, and the tank should be replaced.

Nozzles. The nozzle orifice is an opening with a precise diameter and shape to produce a specific spray pattern and droplet size. With normal use, the orifice diameter will increase, which increases flow rate. Wear changes the shape of the orifice from oval to round, which will change the spray pattern. These alterations occur gradually and may not be noticed.

The replacement schedule for nozzles depends on the amount of use time, the maintenance and cleaning of filters, and the insecticide formulation delivered through the system. Daily spraying with wettable-powder formulations will produce more nozzle wear than spraying with emulsion concentrates.

Replacing nozzles on a 2-year cycle will maintain the intended flow rate, droplet size, and spray pattern.

- *Termiticide application tool tips.* The directional tips on termiticide soil application rods wear with repeated use. Soil particles can abrade the stainless-steel tips. The flow rate and distribution pattern can be altered with long-term use. Soil particles can abrade the threads of the tips to the extent that they can not be tightened to the end of the rod.

Termiticide application tips can be calibrated by using clean water or termiticide dilution in the tank. Maintain a pressure of less than 50 psi at the valve, and release liquid into a 1-gallon container for 30 seconds. If more than 1 gallon is delivered, the nozzle should be replaced. After calibration, return the termiticide dilution to the tank.

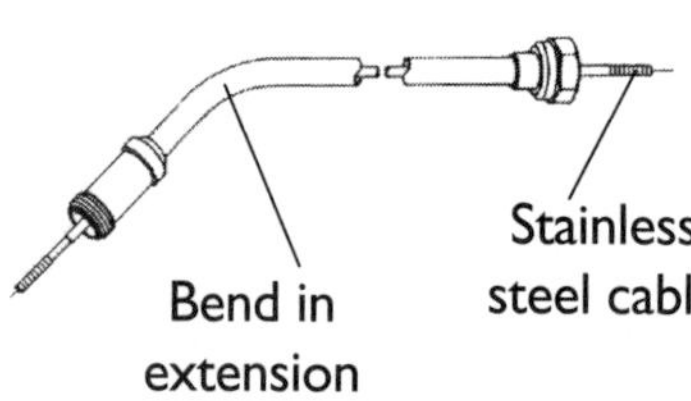

Valve Extensions. Extensions used with tip shut-off valves may wear at the underside of the bend. The plastic-sheathed, stainless-steel cable inside the extension contacts the inner surface at the bend. The surface of the plastic sleeve and wettable-powder particles can rub against and abrade the inside of the extension. Continued use results in a small hole developing on the inside bend in the extension.

Gaskets and O-Rings. The seals in a sprayer are exposed to insecticides, solvents, and constant pressure; they will fail if not regularly replaced. The cracks and composition breakdown in gaskets and O-rings may not be easily seen, but gaskets that seem to suddenly fail have actually been gradually declining. Preventive maintenance by regular replacement of key gaskets can eliminate failure during use.

A stainless steel extension is recommended if continued use of wettable powder insecticides is anticipated.

Environmental Conditions

- *Heat.* Tank sprayers and termite tools can be exposed to hot temperatures when left in service vehicles in the sun. In general, this has little or no influence on the performance of metal-application equipment. Heat is quickly spread across the metal to prevent hot spots and damage, and the expansion coefficients of brass and steel are relatively close.

- *Cold.* When air temperature is below freezing, the water in the siphon tube, hose, and extension may freeze, and the expansion pressure of ice can crack or rupture these parts. Freezing damage can be prevented by relieving pressure in the tank and depressing the valve trigger while elevating the extension to drain liquid back into the tank.

CARE AND MAINTENANCE: TANK SPRAYER

Care and maintenance for daily-use equipment can be divided into monthly inspection, cleaning, and repair; and annual or semi-annual parts replacement.

- *Monthly service* concentrates on inspection of the pump mechanism, valve, hose, and nozzle. Thorough cleaning and flushing of the entire system will prevent build-up of insecticide residue in the tank and hose, and prevent dirt from accumulating in filters.

- *Periodic annual or semi-annual service* concentrates on replacement of worn parts before they fail, as a preventive maintenance procedure. Parts and instructions for routine service are available in manufacturer-prepared kits and manuals. Other replacement components are available from manufacturers as complete assemblies, including pump cylinder and shut-off valve.

Tank

- *Cleaning.* A long-handled, plastic brush with nylon bristles is effective for cleaning stainless-steel and plastic tanks. The end of the brush should be directed at the seam between the bottom and sides of the tank. The majority of pesticide residue will be at this seam.

Liquid remains in the bottom of the tank along the edge and out of reach of the siphon tube. This 1 ounce of liquid dries and creates an insecticide residue in the tank.

Take care when cleaning around the brass siphon tube; bending or damaging the end of this tube will make removal and replacement difficult.

After cleaning, properly dispose of the cleaning liquid and insecticide residue, then flush tank with clean water, and place upside down to dry.

Tank cleaning

- *Pin holes.* Stainless-steel tanks cleaned with chlorine bleach or other chlorine-based products will develop pin-sized holes along the seams. These holes originate from inside the tank and are caused by the corrosive action of chlorine. They cannot be repaired, and the tank should not be used once these holes develop.

Sprayer hoses used to apply wettable-powder formulations, borates, or foam should be flushed daily with clean water.

Hose. The surface of the hose should be inspected for cuts, cracks, burned or scorched areas, and other damage. The junction with the threaded connector should be inspected on sprayers that have been in service for 2 to 3 years. Cracks or weak points at the hose ends may expand with continued use, resulting in leaks or breaks.

Note: Liquid pressure in the hose is equal to that in the tank; a leaking or ruptured hose will result in insecticide exposure and contamination of non-target surfaces.

Pump Mechanism. The pump mechanism provides air pressure in the tank. The pump-cylinder gasket fits tightly around the top of the pump cylinder to seal the pressurized tank. A check valve seals the bottom of the pump cylinder. It opens on the plunger down-stroke as air is forced into the tank, and closes tightly against the bottom of the pump cylinder on the up-stroke of the plunger to prevent liquid from entering the cylinder.

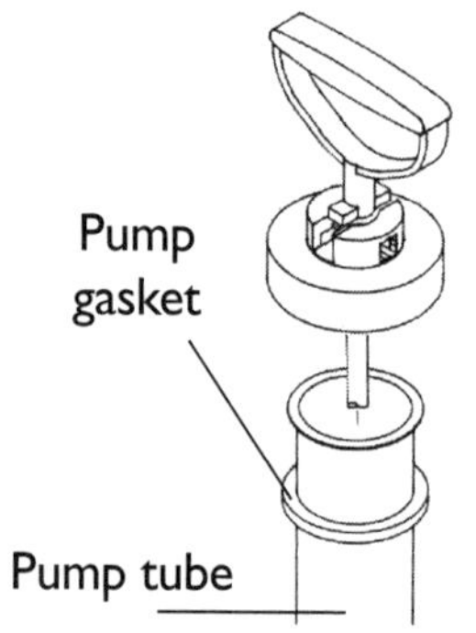

- *Replacing the pump gasket.* The pump gasket is tightly compressed under the tank cap. This gasket should be regularly inspected for cracks, swelling, or other damage that will prevent a proper fit around the cylinder. The pump gasket should be routinely replaced after 1 to 2 years of service.

Swelling of the pump gasket can be detected by counting the number of handle turns needed to attach the pump to the tank. If it is 2 1/2 turns or fewer, the pump gasket is enlarged and must be replaced.

Cracks in the pump cylinder can allow air to enter and exit around the pump handle.

- *Replacing the pump tube check valve.* When the check valve is damaged or not functioning properly, water will enter the pump cylinder and leak out the top during the up-stroke. Replace the check valve every 6 months to 1 year when the tank is in daily use. Clean the base of the cylinder with a nylon brush to provide a smooth contact surface for the check valve.

When replacing the valve, twist while inserting it into the hole for easier insertion. Work the pump several times to be certain the valve is secure before returning the cylinder to the tank.

- *Replacing the pump handle, rod, and spring.* A bent handle or pump rod will result in damage to the pump cylinder and may injure the operator. For tanks that have been in service for many years, it will be necessary to securely hold the pump rod to unscrew the brass handle from the rod. In cases of severe damage or old tanks, it may be necessary to replace the entire pump assembly.

  The pump lock spring may weaken after long-term use and eventually fail to hold the handle in the locked position. It is best to begin at the end of the pump rod and replace the plunger cup and spreader along with the lock spring.

- *Replacing the plunger cup.* The plunger cup should be regularly lubricated with petroleum-type grease—do not use an oil spray. Lubrication will maintain the smooth pumping action and a tight seal with the pump cylinder. The plunger cup should be replaced when the edges are worn or cracked. To replace the plunger cup and spreader plate, remove the nut and washer at the bottom of the pump rod.

- *Replacing the pump cylinder.* The pump cylinder can be damaged with misuse. Dropping the pump mechanism when filling the tank will damage the bottom of the cylinder; hitting the side of the cylinder will create dents, which will reduce the seal of the plunger cup. Severely damaged pump cylinders cannot be repaired and must be replaced.

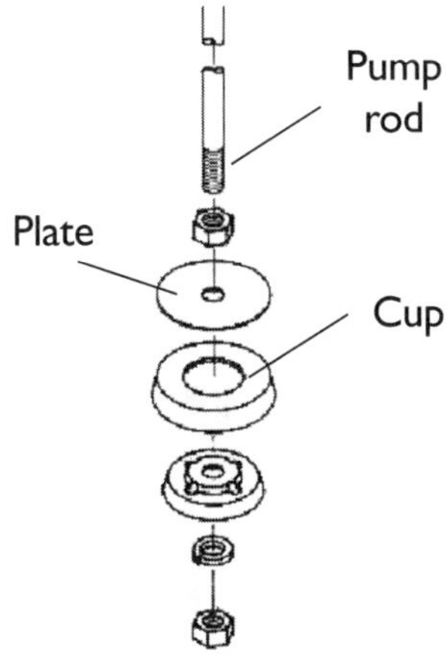

The loss of pumping pressure due to dents in the side of the pump cylinder can be overcome by replacing the plastic plunger with a leather cup. The leather will form a better seal around small dents during the down-stroke of the plunger.

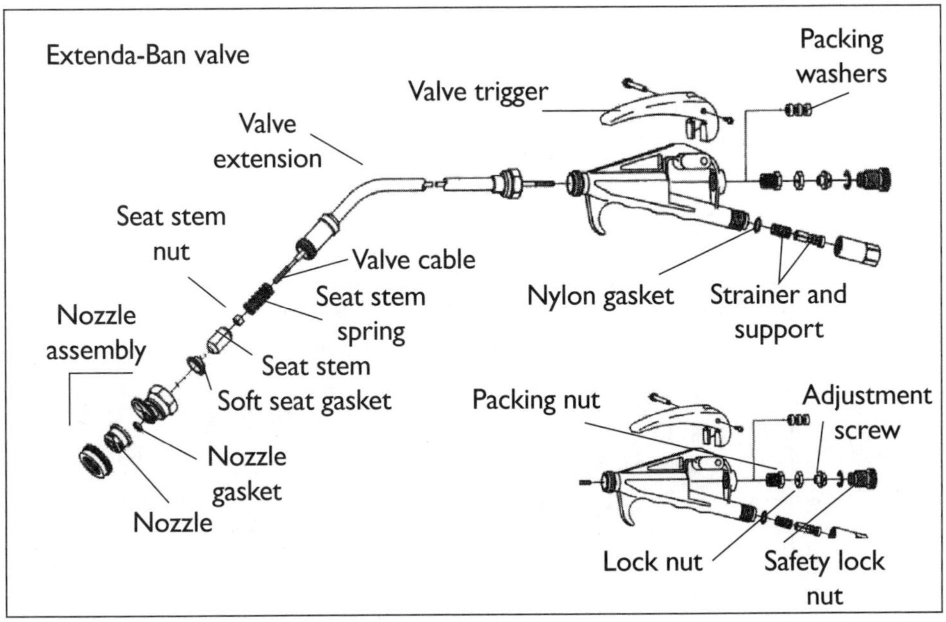

Extenda-Ban Valve. In this valve, the trigger is connected to a stainless-steel cable that extends to the nozzle. The cable inserts from the front through the packing nut at the back. The lock nut and adjustment screw attach to the end of the cable. The safety lock nut is secured by a snap ring. When the lock nut is turned clockwise the trigger is locked.

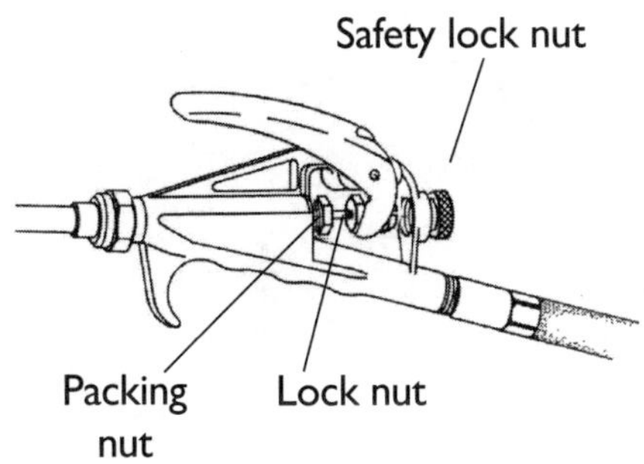

- *Cleaning the strainer.* Liquid from the tank flows through a strainer before entering the valve. The strainer removes dirt from the tank liquid; it must be cleaned regularly with a plastic brush.

- *Cleaning the soft-seat gasket.* The soft-seat gasket should be cleaned with a plastic brush or replaced if it is worn. The gasket is inside a brass housing. Insecticide residue, especially from wettable powder formulations, will accumulate on this gasket and impair the no-drip function of the tip.

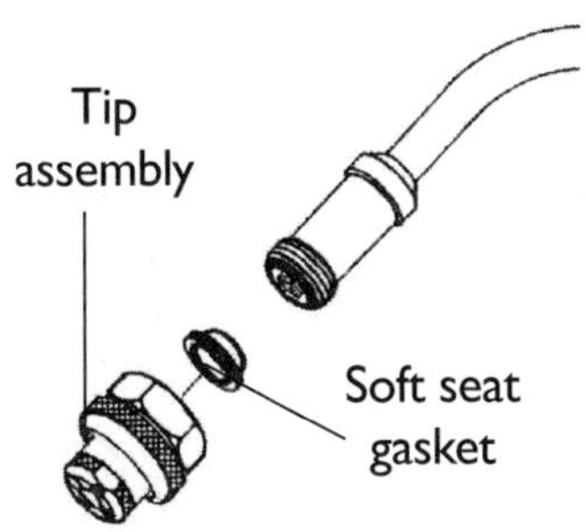

When replacing the gasket, insert it so the *concave side* is toward the tip. To re-attach the tip assembly, depress the valve trigger to retract the cable.

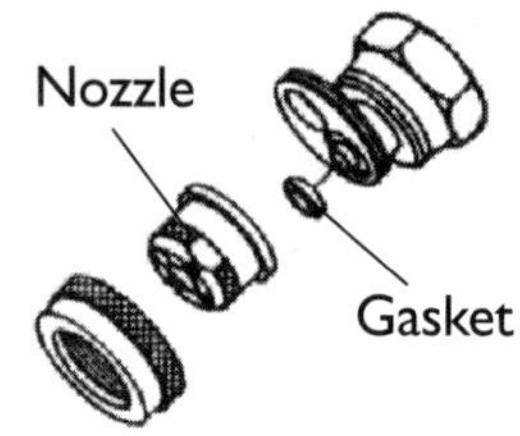

- *Cleaning the nozzle.* The nozzle should be removed and cleaned inside and out with a plastic brush. Nozzles should not be cleaned with metal objects or metal brushes.

- *Replacing the nozzle gasket.* The nozzle gasket seals the base of each orifice, and prevents liquid from leaking at the back of the nozzle. This gasket may become dislodged when the tip assembly is removed to change spray patterns.

Replace nozzles that are damaged or more than 2 years old.

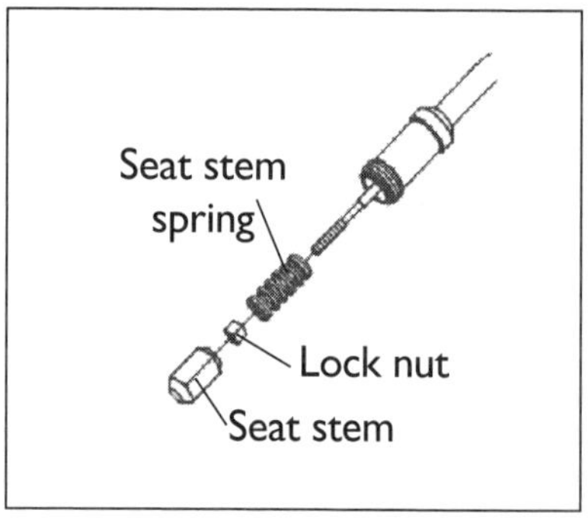

- *Replacing the seat-stem spring.* This spring maintains pressure on the seat stem and prevents drips at the nozzle. To replace:
  - o Remove the nozzle assembly from the end of the extension.
  - o Disconnect the cable at the valve trigger by removing the safety lock nut, then remove the adjustment screw and lock nut.
  - o Push the cable toward the valve until the spring and seat stem are exposed at the end of the extension.
  - o Unscrew the seat stem and lock nut, and remove the spring. Slide the new spring onto the cable, then screw the lock nut onto the end of the cable. Screw the seat stem onto the cable, and hand-tighten it against the lock nut. *The lock nut must be positioned about **4 threads** down from the start of the threads.*

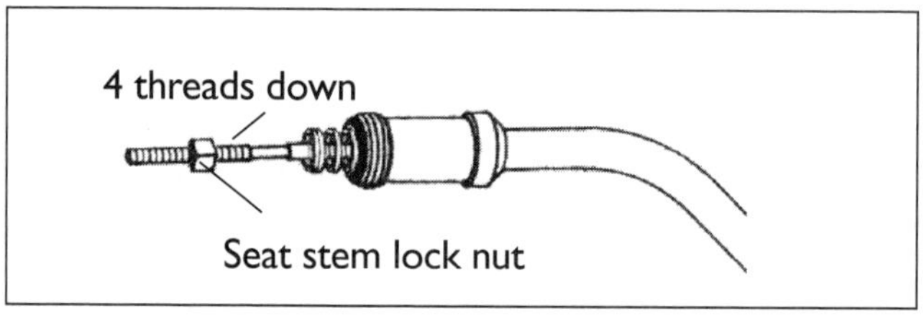

  - o Pull the cable from the back of the valve until only the tip of the seat stem extends out of the end of the extension. Re-connect the cable at the back of the valve: attach the lock nut, followed by the adjustment screw, then position the trigger between these two.
  - o Re-insert the safety lock nut and snap ring.
  - o Re-attach the tip assembly and the soft seat gasket to the end of the extension. *Position the soft-seat gasket with the concave side facing toward the tip. If reversed, the seat stem will not seal against the gasket, and the tip will leak.*
  - o Tighten the tip assembly to the end of the extension by using two wrenches; this will prevent twisting and damaging the extension.

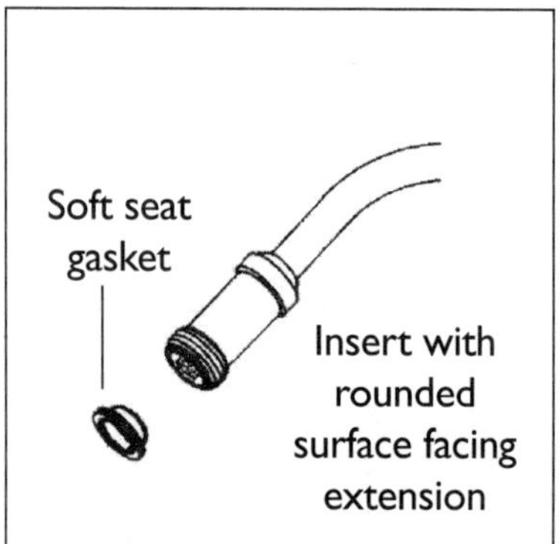

- *Replacing the valve cable.* Replacing the valve cable requires disconnecting parts at the end of the extension and at the back of the valve.
  - o The valve cable must be removed through the back of the valve; it can not be removed through the end of the extension. The long end of the cable will not pass through the curved end of the extension.
  - o Insert the long end of the replacement cable into the end of the extension and into the front of the valve. The cable should pass through the valve and through the packing nut.

It may be necessary to remove the packing nut, then slide it over the end of the cable once the cable has been inserted the length of the extension.

- o Replace the parts at the front end of the extension, and re-connect the parts to the other end of the cable.

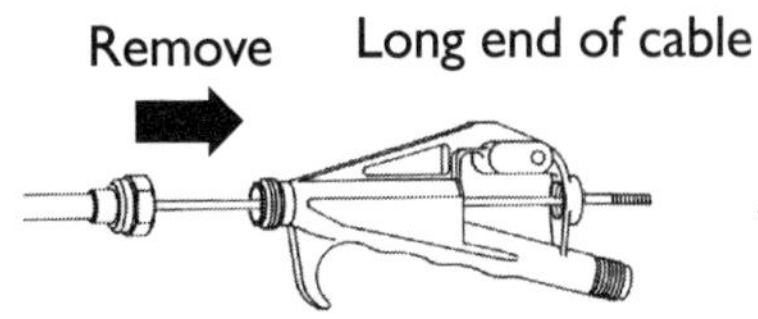

Removing valve cable

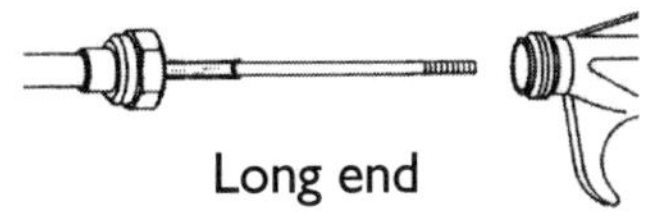

- *Replacing the valve packing washers.* The Extenda-Ban valve has washers that form a seal around the cable. The packing washers are inside where the cable exits the back of the valve; they are held in place by the packing nut. The packing nut can be tightened to stop minor leaks around the cable, but the washers must be replaced when the leak persists.
  - o To replace the packing washers, the valve cable must be disconnected from the back of the valve, and the packing nut removed. Referring to diagram below:

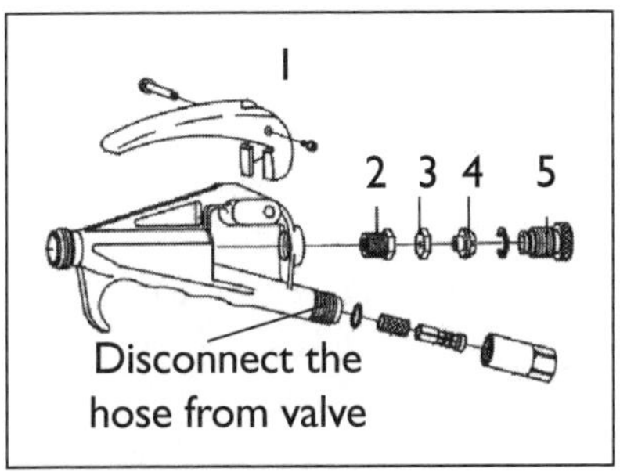

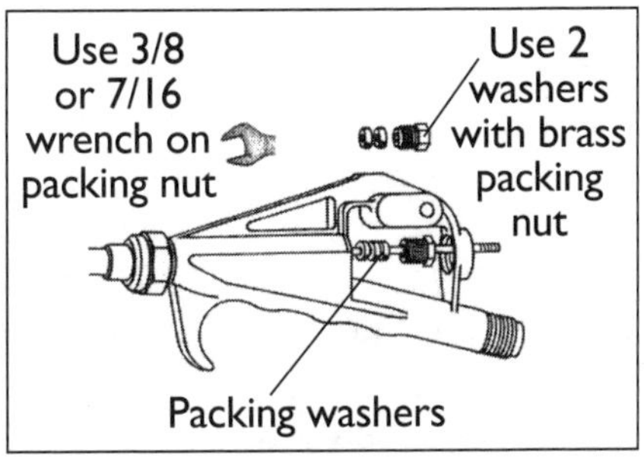

1. Remove the trigger screws and remove the trigger (1).
2. Remove the safety lock nut (5); then remove the adjustment screw (4) and lock nut (3).
3. Disconnect the extension from the front of the valve, and slowly pull the cable through the front of the valve.
4. Remove the packing nut from the valve.

- o The white packing washers are in front of the packing nut. Remove the washers carefully, without damaging the internal threads in the valve.
- o Insert the end of the cable into the front and through the back of the valve. Slide the washers onto the cable, followed by the packing nut. Slightly tighten the nut; further tightening will be done once the valve is re-assembled.
- o Reconnect the trigger and parts at the end of the cable, and re-attach the extension to the valve.
- o Pressurize the tank and inspect the valve for any leaking around the packing nut; tighten the nut only enough to eliminate any leak. Excess tightening may damage the packing washers and reduce their service life.

- *Replacing the hose and siphon tube.* In most compressed-air sprayers, the hose to the valve is connected to the siphon tube at the top of the tank. The hose may wear at this connection after prolonged use, or it may be damaged while in the service vehicle. A worn or damaged hose must be replaced immediately.
  - o The hose is connected to the tank with an adapter that joins the top siphon tube to the end of the hose. Remove the hose from the adapter, then remove the brass adapter. The expanded top of the brass siphon tube will be visible in the top of the tank. It may be necessary to dislodge the tube from inside the tank.

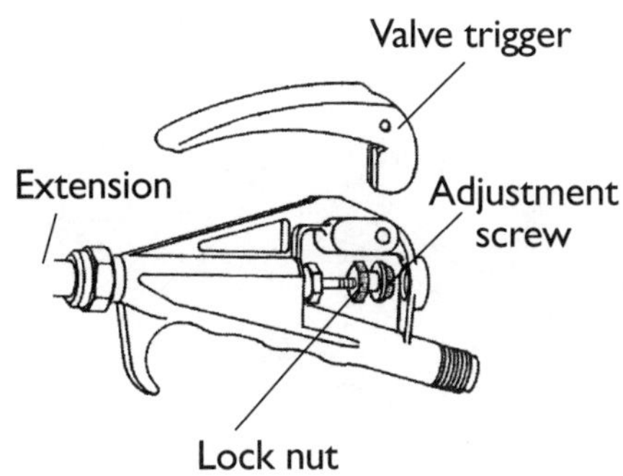

o Remove the hose from the valve, and retain the nylon washer that seals the junction of the hose and valve.

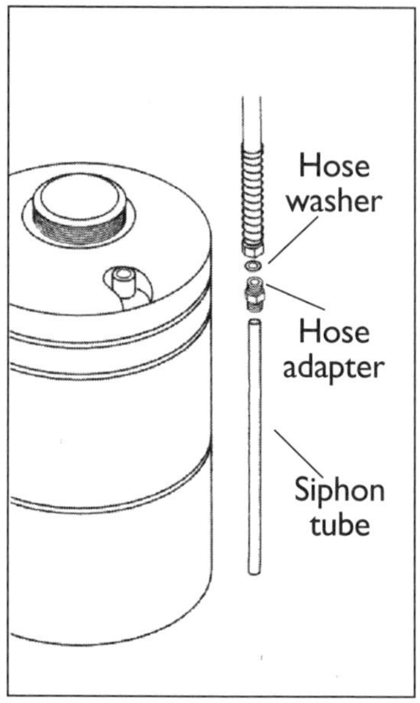

When attaching the hose to the tank hose adapter, do not over tighten the connection. This will compress the soft nylon hose washer so that it obstructs the end of the hose and prevents the flow of liquid. The hose washer must be replaced when damaged.

- *Repairing T-Jet and XR-Valve.* After long service it may be necessary to replace the valve body or the components of the shut-off mechanism. Manufacturer repair kits usually contain all the valve components for a complete overhaul or maintenance. Leaks originating at the base of the valve may be stopped by tightening the nut of the base cap at the bottom of the valve. Continued leaks from the bottom or top of the valve require disassembly and replacement of some or all of the internal components.
  - o The base cap must be removed from the bottom of the valve body. This brass nut is tightly secured and must be removed carefully so as to not damage the valve body. There is a copper gasket on the base cap. It should remain in place, or be replaced with other components of the valve.
  - o The valve spring may become weakened after years of use. This spring fits into the top of the base cap and into the bottom of the valve stem.
  - o There are two components on the valve stem: a valve washer and an O-ring. These should be replaced whenever the valve is repaired, as leaks generally originate from these parts.

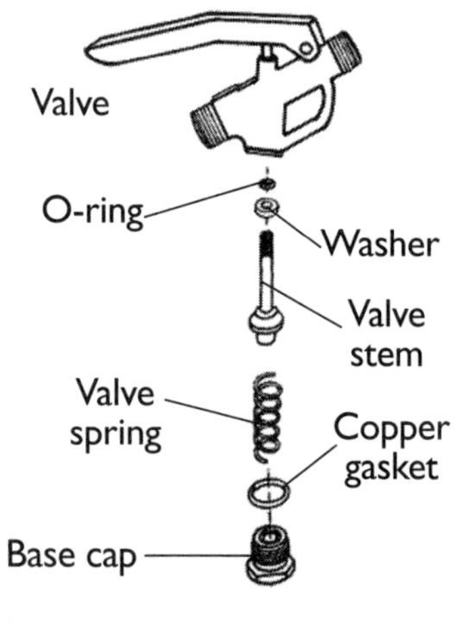

- *Cleaning the check valve and filter.* The check valve is at the end of the extension, behind the nozzle, and is usually combined with a screen filter. The filter will become clogged with particulate matter from the spray liquid and must be regularly cleaned.
  - o Remove the tip assembly from the end of the extension; use two wrenches to prevent twisting and damaging the extension.
  - o Clean the filter screen with a nylon brush, then replace and re-attach the tip.
  - o If the check valve is replaced, the new check valve must have the same pressure rating as the old valve. A higher rating may limit normal spraying, and a lower rating may not provide the anti-drip feature.

TANK SPRAYER PROBLEMS AND REPAIRS

The mechanical problems that occur with tank sprayers are usually the result of a poor repair and maintenance schedule. These problems can be corrected by cleaning or gasket replacement. The leaks that occur at the valve or the nozzle are usually due to the failure of a gasket or washer.

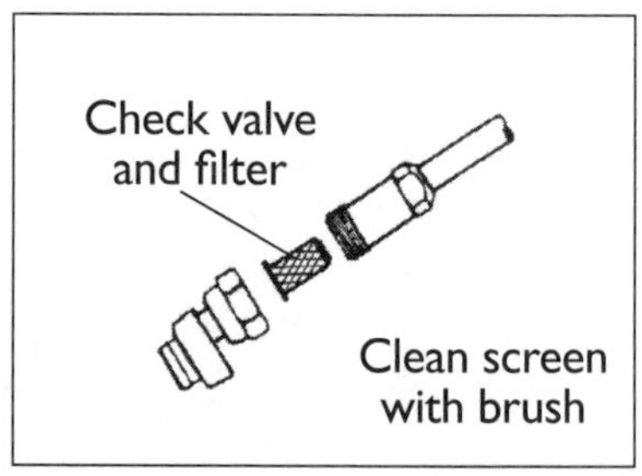

Manufacturers provide detailed parts lists,

order numbers, and instructions for product repair in manuals or on their websites. Repair kits contain the gaskets, washers, springs, and other parts necessary to correct common problems. Manufacturers also provide complete assemblies of original valves and extensions, and pump mechanisms.

Tank

- *Problem.* Leaks along the welded seams or small pin-hole leaks along the side and bottom seam of the tank.
  Repair. The pin-hole leaks at the seam cannot be repaired, the tank must be replaced.

- *Problem.* Pump handle is loose in the tank cap; it does not remain locked when turned into the tank top.
  Repair. The pump lock spring is weakened and does not provide the pressure to lock the handle. Replace the lock spring.

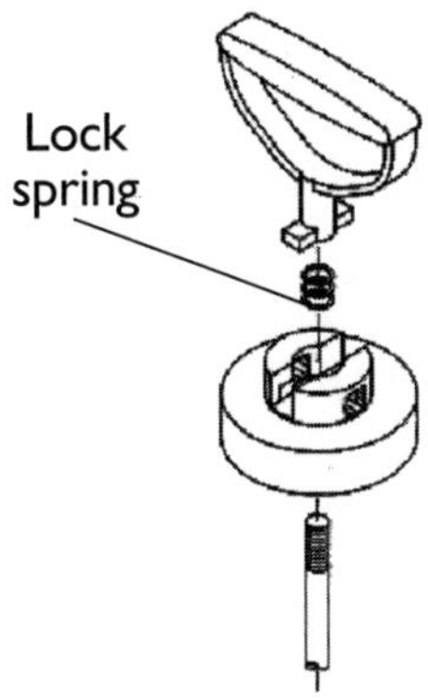

Tank Pressure

- *Problem.* Tank does not hold pressure; pump is working properly but little or no air pressure remains in tank.
  Repair. Inspect pump gasket at the top of the pump cylinder; it may be cracked or damaged and not forming a seal with the tank cap. Replace the pump tube gasket; it should fit tightly around the top of the pump cylinder.

- *Problem.* Tank is not getting pressurized; the pump is working but no pressure builds in the tank.
  Repair. Apply grease to the edges of the plastic cup and to the inside of the pump cylinder. Replace plunger cup. If pump cylinder is dented and not forming a seal with the plunger cup, replace the plastic cup with a leather cup.

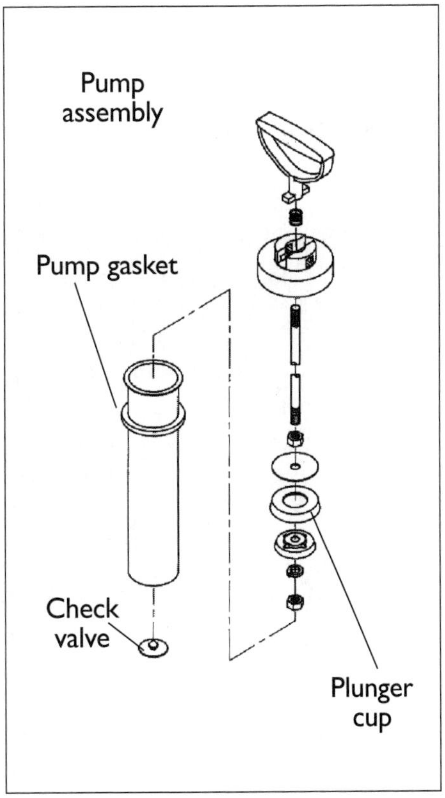

- *Problem.* Pump cylinder fills with liquid, and pumping is difficult; the tank is not getting pressurized.
  Repair. Clean residue from the bottom of the pump tube. Replace the check valve at the bottom of the pump cylinder. Be certain that it is secure in the bottom of the cylinder before returning the pump to the tank.

Spraying Failure

- *Problem.* There is little or no spray coming from the nozzle, although the tank holds pressure. This may occur in a new sprayer that was just assembled.
  Repair. Inspect the hose washer in the end of the hose. It may have been crushed

when the hose was attached and is blocking the flow of liquid. Replace the hose washer. Do not over tighten the hose to the tank, this will damage the hose gasket.

- *Problem.* There is only a small amount of spray or no spray from the nozzle; there is no problem with the pump, and the tank holds pressure. This may occur in a sprayer that is several years old.
  Repair: Siphon tube may be clogged at the bottom inside the tank, or the tube may be cracked. Siphon tube must be replaced if cracked. To do so:
  1. Remove the hose from the valve, and clean the strainer that inserts into the valve.
  2. Remove the nozzle and clean with a nylon brush.
  3. Remove the nozzle assembly from the end of the extension, and clean or replace the soft seat gasket, which is located between the nozzle assembly and the end of the extension.

- *Problem.* Nozzle is not delivering a consistent fan-spray pattern. It is distorted to one side, sprays only in the middle, or is otherwise uneven.
  Repair. Replace the nozzle. The nozzle is probably more than 2 years old and has become worn, or the nozzle has been cleaned with a metal object and damaged.

Leaking at Valve, Extension, Nozzle, or Hose Connection

- *Problem.* Valve is leaking around the packing nut; origin of the leak may not be visible but the handle is wet.
  Repair. Tighten the packing nut one-quarter turn; use a 3/8 or 7/16 wrench. If this does not stop the leak, the packing washers must be replaced.

- *Problem.* Valve is leaking at the connection of the hose to the back of the valve; the hose is wet.
  Repair. Tighten hose connection to the valve one-quarter turn; if leak continues, replace the nylon gasket between the hose and valve.

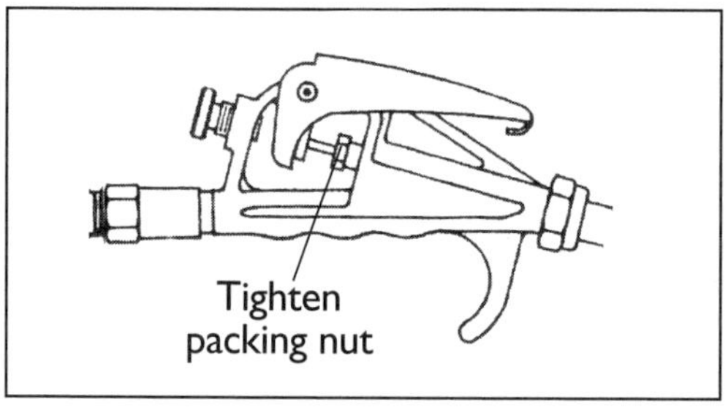

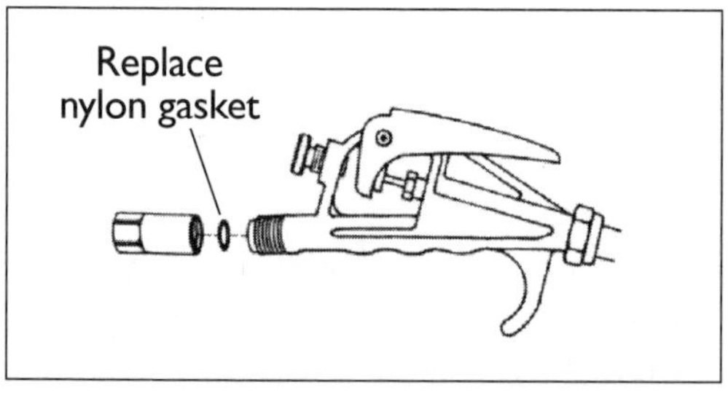

- *Problem.* Nozzle is leaking at the back when the valve is open, or there is a constant drip from the tip; the spray pattern is weak or incomplete.
  Repair. Remove the nozzle, and inspect the nozzle gasket; clean the gasket, or replace missing gasket.

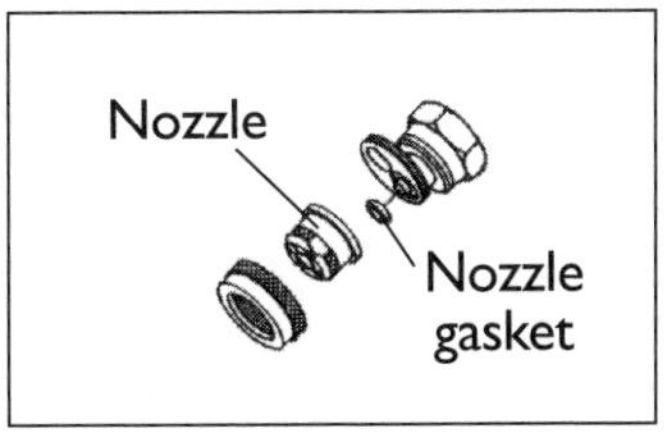

- *Problem.* Leaking around the connection of the hose to the tank; leak is noticeable when the tank is fully pressurized.
  Repair. Replace the hose washer inside the end of the hose; do not over tighten when reconnecting hose.

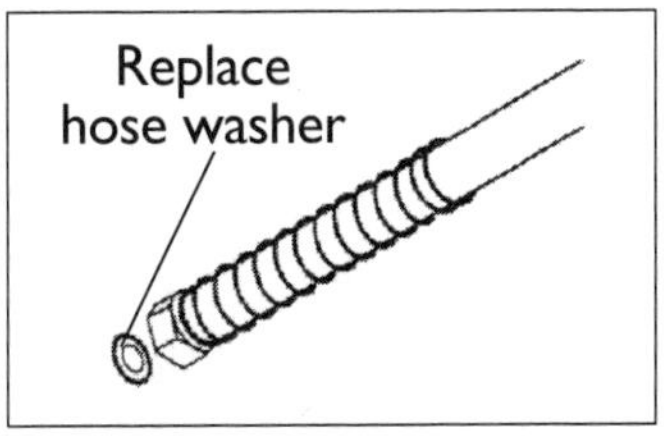

CHAPTER 19

PESTICIDE APPLICATION SAFETY

The risk of exposure to pesticides begins with mixing of the concentrate and extends through application to cleaning of the equipment. Within these procedures, there is potential for harmful insecticide exposure for applicators, non-target organisms, and the immediate environment. Pesticide product labels address these risks and the responsibilities through precautionary statements. They explain the application hazards to humans, domestic animals, and non-target organisms, including honey bees, fish, and other aquatic life. In addition, product labels include precautions for standing or flowing water and run-off, as well as physical and chemical hazards, including electrical contact and combustibility. Labels also provide first-aid instructions in case the insecticide is swallowed, contacts skin or eyes, or is inhaled.

ROUTES OF PESTICIDE EXPOSURE

Pesticide exposure and routes of entry into the body of people or pets can occur by absorption through the skin (dermal), the mouth (oral), or inhalation of small, 5- to 10-micron particles or droplets into the lungs (respiratory). All of these can occur during routine preparation and application of household and structural insecticides.

Dermal. Dermal exposure results in absorption immediately after a pesticide contacts skin or eyes. Absorption continues for as long as the pesticide remains on the surface. The rate at which dermal absorption occurs is different for different regions of the body.

Using the forearm absorption rate of 1.0 for comparison, absorption rates for other body regions, are:

- palm of the hand, 1.3
- ball of foot, 1.6
- abdomen, 2.1
- scalp, 3.7
- forehead, 4.2
- genital area, 11.8

Exposure may also occur through the transfer of insecticide or residue from one part of the body to another. For example, small children may crawl on or otherwise contact treated surfaces by hand and subsequently put their hands in their mouths.

Dermal exposure of dogs, cats, and reptiles may occur when these animals walk on treated surfaces before the spray has dried. Insecticide labels recommend that treated surfaces be allowed to dry before pets are allowed in the treatment area.

Oral and Respiratory. Oral exposure often occurs during pesticide mixing and spraying or cleaning of equipment. Ingestion can occur accidentally when backsplash from overhead spraying strikes the face. Personal protection equipment, such as goggles or glasses, are often required by the label and should be routinely worn during mixing and spraying. Wearing of gloves during mixing and spraying can prevent insecticide exposure and transference of residue to the mouth.

Oral exposure can also occur when pesticides are placed in an unlabeled jar, bottle, or food container.

Some insecticides can cause serious injury to nose, throat, and lung tissue when inhaled; low exposure rates can cause temporary irritation and respiratory problems for people and pets. Respiratory exposure is especially hazardous because the toxic material is usually too small to see, and pesticide particles can be rapidly absorbed into the blood stream through the lungs.

Using the mouth to clear a spray nozzle or hose can result in exposure to diluted or concentrated residue deposits.

Droplets below 10 microns can be inhaled and are absorbed by the lung tissue. From there, the insecticide passes into the bloodstream.

Exposure to airborne insecticide particles can easily occur during mixing of dry formulations, such as wettable powders, wettable granules, dusts, and even granular insecticides. Household pets may be exposed to airborne insecticides immediately after indoor spraying.

Particles measuring 1 micron or less, such as those of vapors and smoke, can be inhaled and exhaled through the lungs.

The small droplets generated by fan-spray nozzles can remain in the air and may be a threat to caged birds, reptiles, and their exposed food in the treated room.

- *Service vehicles.* There is potential exposure to trapped insecticide vapors in pest control service vehicles, such as pick-up trucks, vans, and sedans. Especially prone to this type of exposure are vehicles in which application equipment, such as compressed-air sprayers, are stored in the cab with the driver or in the back of a van or sedan.

Fall time for 50-micron droplets at the edge of a fan-spray pattern is less than 1 minute, but indoor air movement may move droplets to non-target sites. The hazard from drifting spray droplets is low when a coarse fan-spray application is used.

There is significantly more insecticide in the air of a truck cab containing a sprayer than there is when sprayers are kept in the bed of the truck or the back of a van. Spilled insecticide inside the cab will also contaminate the air space.

- *Treated rooms.* Exposure to insecticide in the air of treated rooms immediately after application appears to be dependent on the air movement in the room, such as by ceiling fan, floor fan, and air conditioning, and not on the type of spray application. In general, when there is no forced-air flow in the treated room, there is less insecticide in the air immediately after and 24 hours after application. Even slight air currents can move insecticide from application sites into the ambient air.

PERSONAL PROTECTIVE EQUIPMENT

In general, all technicians mixing, loading, and applying pesticides should wear a long-sleeved shirt and long pants, socks, shoes, and usually chemical-resistant and water-proof gloves. Applicators may be instructed by label directions to wear protective eyewear when applying liquid termiticides.

Laundering. For most service technicians, the standard personal protective equipment of shirt, pants, shoes, and socks are laundered and reused throughout the year. While laundering is the most common decontamination method, it is rarely complete, and residues remain. Insecticide residue on daily worn clothing may be on the fabric surface or matrix. Matrix contamination occurs when a chemical permeates or penetrates into the fabric or fiber. It is not visually detectable and not readily removed by laundering.

Gloves and Boots. Product labels may require that applicators use water-proof or chemical-resistant gloves during mixing and application. However, service technicians may also want to consider wearing protective gloves during routine insecticide application. Glove material may be specified by the product label, such as barrier laminate, butyl rubber, nitrile rubber, neoprene rubber, polyethylene, or polyvinyl chloride. The thickness of protective gloves is generally14 mil.

Use of chemical-resistant gloves is generally linked to the solvent system of the formulation, not to the active ingredient. Different formulations of the same active ingredient may require different chemical-resistant gloves or other protective equipment.

Chemically resistant boots may be required by the label when applying some liquid insecticides. The material for these boots is the same as that for gloves. High-top boots can provide protection from liquid termiticides, especially during exterior perimeter application.

Eye Protection. Termiticide product labels typically require that applicators use protective eyewear when working in non-ventilated spaces and when delivering liquids by rodding and sub-slab injection. The objective is to shield eyes from small droplets during spraying and from splashback during injection. The product label on fumigants may require that the operator wear goggles or a full-face shield when handling or introducing the fumigant into a site. Goggles are defined as eye protection that fits tightly to the face around the eyes. The objective is to prevent exposure to the gas in the ambient air.

To be rated as safety glasses, both frame and lenses must conform to the American National Standards Institute standard: ANSI Z87.1-2003.

Safety glasses are usually made of polycarbonate plastic. They generally include anti-scratch and anti-fog features; and some have ultraviolet coatings to improve outdoor vision. Glasses are also available with prescription lenses. Basic fitting requirements include that they be comfortable; fit snugly; and are durable, easily cleanable, and capable of being disinfected. Fitting at the forehead and temples is important for complete protection.

- *Ultraviolet light.* The blacklight (BL) bulbs used in insect light traps generally emit energy in the 350 to 400 nm region of the light spectrum. That band is within the "UV-A" region of the spectrum. BL bulbs emit both visible light and ultraviolet light. Eye protection is not necessary when working with these bulbs. The blacklight blue (BLB) bulbs that are used in specialty or effect lighting are made with special glass that filters out most of the visible light, while allowing the UV-A light to pass through. Blacklight blue bulbs are classified as Risk Group 1 (ANSI/IESNA RP-27.3-96). This category is referred to as low risk with normal use; but prolonged exposure should include eye and skin protection.

Ear Protection. Hearing loss can result from short- or long-term exposure to loud noise. The most common type of permanent hearing loss in the workplace develops from repeated exposure to loud sounds over an extended period. The U.S. Occupational Health and Safety Administration (OSHA) requires that employers implement hearing conservation programs for employees if the noise level of the workplace is 85 dB or higher for an 8-hour period.

Service technicians may need hearing protection when using some equipment for termite and pest control. Examples include technicians who use electric hammer drills to treat beneath concrete slabs, or those who use certain types of thermal (gas-powered) fogging machines. The use time of these tools may be limited to only a few hours each work day, but ear protectors should be available when needed. The most common types are: expandable foam plugs, pre-molded plugs, ear-canal caps, and earmuffs.

The use of ear plugs does not make it difficult to hear others talking. Ear plugs reduce noise most efficiently in the low frequencies, below the speech range, and make speech more easily heard.

Head Protection. Hard hats or bump caps are necessary for service technicians working where there is danger from impact or falling objects. The American National Standards Institute (ANSI) has established specifications for helmets to protect against impact and penetration of falling objects. Hard hats require a hard outer shell and a shock-absorbent lining. They are divided into three categories: Class A, for general service to provide impact protection but limited voltage protection; Class B, for protection against falling objects and high-voltage shock and burns; and Class C, for limited protection through lightweight helmets.

Pest-control and termite-control technicians may require only a Class C hard hat, bump cap, or baseball-style cap with molded-plastic, inner skull cap. These caps provide basic protection when working in building crawlspaces, basements, and attics.

Particulate filters are designed to reduce the velocity of air passing through the filter.

Respirators. Pesticide formulations, such as wettable powders, require applicators to wear respirators when treating overhead surfaces. Full or half-face respirators use a filter which acts passively on air inhaled by the wearer. The half-face masks are typically a single-piece design that traps and removes particulates from the air stream. The most common of these facemasks is the disposable, white N-95. The entire unit is discarded after extended or single use, depending on the contaminant.

N filters are not resistant to oils or solvents; R filters are oil and solvent resistant for up to 8 hours; P filters are oil and solvent proof.

The HEPA (High Efficiency Particulate Air) designation has been replaced with N-100, R-100, and P-100 filters. The new type 100 respirators are designated as "HE" (high efficiency). They are capable of removing particles in the air down to 0.3 microns. There are also 95- and 99-series filters, but these provide less protection than the type 100 respirators. The 100 Series protects the user from particulates such as viruses, bacteria, and animal and insect dander.

N-100 filter provides protection from fine particles when there is no oil or solvent in the formulation or as an ingredient in the particles. It also provides protection against hantavirus, as can be transmitted from rodent droppings.

- *Half-facepiece respirators.* These respirators cover the wearer's nose and mouth. Because inhalation creates a slight negative pressure inside the facepiece, these devices are also called negative-pressure respirators. During inhalation, contaminated air enters the facepiece at gaps between the material and the face. A complete face-to-facepiece seal is essential for good protection.

A tight seal at the nose is difficult, and it is the most frequent location of leaks. Facial hair, even stubble, also prevents a complete seal.

In dusty conditions, repeated exposure of the eyes to dust increases the risk for injury and disease. Most dust particles entering a person's eyes will be washed out by tears, but some can be retained, particularly within the margin of the upper eyelid. These particles can become embedded in the surface of the cornea, where they cause reddening of the surface and sometimes infection.

▶ Rubber or elastic respirators should be limited to low-risk jobs for protection against dirt or insecticide-particle inhalation.

Half-facepiece respirators are inadequate for working and applying insecticides in dusty conditions. Eyecup goggles are not airtight and do not completely prevent dust exposure.

- *Canister respirators.* These respirators are typically designed as a half mask, which covers the mouth and nose of the user. Full-face masks cover the eyes in addition to a specific fitting for the mouth and nose. This group of respirators uses a small canister to protect the wearer from breathing in particulate matter, gases, and vapors. Disposable respirators and those with replaceable filters can be used until they are difficult to breathe through or damaged. The respirator cartridge is a metal or plastic case containing activated charcoal.

  Canister filters are usually coded by the type of matter they filter, such as gas and vapors (A=organic, B=inorganic, E=acidic) or particulate (P=particulate, usually including bacteria); and for capacity (1=low, 2=medium, 3=high). For example: a canister coded P3 is designed for solid and liquid particulates and has a high capacity; a canister coded A2 P3 combines organic gas and airborne particulate matter; and a canister coded B2 P3 would filter inorganic gasses and airborne particulate matter. Canisters for half- and full-face masks are available in a variety of A, B, and P combinations and filtering capacities.

Protective Clothing. Cotton and cotton-blend fabrics are the standard washable fabrics for work uniforms. Cotton twill and poplin may retain larger quantities of pesticide solution within fabric spaces rather than transport it to undergarments and skin. Regardless of the type of fabric used in the clothing worn by applicators, protective fabric treatments can increase resistance to pesticide liquids.

▶ Fabric spray starch acts as a pesticide trap by binding with pesticide to keep it on the fabric and away from underlying skin.

Specialized fabric protectants sprayed on work clothes can reduce by 90% the amount of pesticide absorbed. Most fabric protectants remain effective through three washing.

- *Shirts and pants.* The potential exposure of applicator arms and legs during insecticide application is addressed in the Personal Protective Equipment (PPE) required by the label. Long-sleeved shirts, long pants, socks, and shoes are standard. The areas covered by this clothing are most likely to receive insecticide drift during liquid and dust application.

- *Coveralls.* Whole-body coveralls are available for maximum protection against chemicals and airborne particulate matter. The basic covering material varies from polypropylene for low-hazard protection to polyethylene with additional coatings for high-hazard protection. Tyvek™ is a brand of high-density polyethylene fibers; it is a nonwoven material of 0.5 to 10 micron fibers. Water vapor can pass through it, but not liquid water. Coveralls of this material are commonly used when low hazard protection is required.

- *Bee and wasp protection.* A hat and sturdy veil or a combination of hat, veil, and jacket can provide adequate protection from most stinging insects. However, full head-to-ankle protection may be required when working with aggressive strains of honeybees and yellowjackets. The fabric in this protective clothing is usually polyester cotton twill, but non-woven polyethylene material may be used in coveralls. The veil fits over a hard or flexible hat. Long-cuffed leather, rubber, or nitrile gloves are required to protect hands.

- *Laundering work clothes.* Clothing worn during insecticide application should be laundered on a regular basis. Residue on fabric can accumulate with repeated wearing, and removal is less complete as the concentration increases. Because insecticide residue can transfer to other clothing during washing, it is necessary to wash contaminated clothing separately. Pre-soaking clothing in hot water prior to washing, then washing in hot water at 140°F provides the optimal conditions for removing insecticide residue.

PESTICIDE TOXICITY

Pesticides can be classified according to their toxicity, and toxicity can be used to limit the use of specific formulations to trained applicators. The objective of the classification is to identify active ingredients or formulations that pose a threat to humans or the environment and safeguard their use. Applicators should be aware of the hazard associated with exposure to a pesticide, as well as the toxicity of the chemical.

Acute and Chronic Toxicity. Acute toxicity refers to the effects from a single or repeated exposure to a pesticide over a short time. There are signs and symptoms associated with acute poisonings. A pesticide with a high acute toxicity characteristic can be dangerous if even a small amount is absorbed. Chronic toxicity refers to the effects of long-term or repeated low-level exposures to a toxic substance. The effects of chronic exposure may not appear immediately, rather they may take months or years to become evident.

Measuring Toxicity. Toxicity is measured in milligrams of active ingredient for each kilogram of body weight (=parts per million). This is measured as the dose needed to kill 50 percent of a sample of test animals in 24 hours, and is referred to as the LD_{50} dose.

A pesticide with a low LD_{50} is more toxic than a pesticide with a high number because it takes less of the material to kill half the test animals.

Signal Words. A pesticide product label typically has one of three signal words that indicate the degree of toxicity of the active ingredient. The specific signal word is intended to indicate the degree of potential risk to the applicator, not the pest control efficacy of the product. The pesticide signal words and relative acute toxicities are:

- *Danger* – Highly toxic, oral lethal dose is 1 to 5 ml.
- *Warning* – Moderately toxic, oral lethal dose is 5 to 15 ml.
- *Caution* – Low toxicity, oral lethal dose is 15 to 473 ml.

Some neonicotinoid insecticides have "Reduced Risk" EPA status, and because of their limited or lack of toxicity to vertebrates, there are no signal words or warnings on the product label.

MATERIAL SAFETY DATA SHEETS

Material Safety Data Sheets (MSDS) provide information on the hazards of the active ingredients of pesticide products and first-aid measures. These documents are generally available for all registered pesticides. Information contained in the MSDS includes manufacturer identification, chemical properties of the active ingredient, toxicological data, hazard identification, fire-fighting measures, handling, storage, disposal considerations, and ecological information.

Product labels typically include a section recommending procedures for accidental release, spills, or leaks.

CONTAINER DISPOSAL

Triple rinsing is the standard procedure prior to the disposal of liquid insecticide containers. At this level, the pesticide containers are usually considered to be acceptable for disposal with municipal waste without threat of contamination. Rinsing empty containers is often difficult, particularly when the original insecticide was a concentrated formulation with high viscosity. In some SC formulations, the particulate solid may adhere to the sides and bottom of the container.

PESTICIDE ACCIDENT HOTLINE (CHEMTREC)

CHEMTREC is a public service hotline for fire fighters, law enforcement officials, and other emergency responders for obtaining information for emergency incidents involving pesticides and other hazardous materials. It operates 24 hours a day, 7 days a week in the U. S. at 800-262-8200.

INSECTICIDE SPILL CONTROL PROCEDURES

Interior household and exterior perimeter applications typically require small amounts of diluted insecticide in tank or backpack sprayers. There is limited potential for insecticide exposure or contamination from a liquid spill from these tools. Termite control with liquid insecticide requires large amounts of insecticide for pre- and post-construction applications. An accidental spill from a large tank of diluted termiticide may contaminate soil and nearby individuals.

In case of a spill:

1. *Equipment and operations.* The first response to a spill is to limit the danger by stopping the application process. When a spill or accidental treatment occurs, pumps and other tools should be turned off. If the spilled insecticide is flammable, all pilot lights and gasoline engines should be shut off.
2. *Control and contain the spill.* Stop further leakage from the container to prevent it from spreading to storm drains, wells, and waterways. An insecticide spill should not be flushed with water, since this may further spread the insecticide. Spill-control kits contain absorbent material for liquid insecticides. Other materials that may be available on-site include sawdust or cat litter.
3. *First aid.* Follow the product-label directions in providing first-aid assistance. If excess insecticide contacts the skin, wash the affected area with soap and water immediately. Clothing that is wet with insecticide should be removed, placed in a marked plastic bag, and handled according to label directions. The occupants of buildings should be informed of the spill and kept clear until it has been cleaned and the area decontaminated.
4. *Clean-up.* Consult the manufacturer for clean-up procedures in situations in which there is a large amount of material spilled. Using detergent and water will remove practically all detectable surface residue. When carpet padding has become wet from a spill, the affected area will have to be removed and replaced. The sub-flooring may need to be washed with a detergent solution. Spills on linoleum, concrete, and other smooth and finished surfaces can be absorbed with paper towels; the surfaces can then be cleaned with detergent.

DECONTAMINATION OF HOUSEHOLD SURFACES

Indoor application of insecticides may lead to contamination of surfaces or other non-target objects. Considering the wide range of active ingredients and formulations, it is difficult to establish cleaning procedures, materials, and methods for household insecticides. Surfaces treated during typical pest control operations have different absorption and desorption characteristics. The most practical method for decontaminating surfaces such as polyvinyl floor coverings, glazed ceramic tile, Formica counter-tops, and polyamide carpet is to use common cleaners and scrubbers.

Non-porous Surfaces. There is little or no absorbency of insecticides on ceramic tile, plastic Formica laminate, polyvinyl floor covering, and painted wood. Insecticide residue remains at the surface and can be removed with the combination of multi-purpose or alcohol-based cleaners and a power washer.

Pile Carpet. The process of decontamination of carpeting includes initial vacuum cleaning with a rotating-brush vacuum, followed by the use of a spray-extraction cleaner to apply a carpet detergent. The detergent or cleaning solution is applied to the carpet and allowed to remain for about 5 minutes before making a second application and simultaneous vacuuming. The procedure of simultaneous cleaner application and vacuum removal must be repeated several times, depending on the amount of insecticide in the carpet. After the carpet is dry, it can be vacuumed again.

Polyvinyl Chloride Floor Covering. Removal of insecticide and residue in this thermoplastic polymer can be done with a basic alkaline cleaner (pH=12.1). The cleaning solution can be applied to the floor with a cloth, and after about 10 minutes the floor should be scrubbed with a commercial scrubbing vacuum cleaner. This procedure may be repeated several times and can achieve 99% reduction in the active ingredient.

CHAPTER 20

MEASUREMENTS AND CALCULATIONS

Standards of measurement used in insecticide label directions, equipment specifications, and other aspects of pest control may be in inch-pound units or in metric. Converting between these two systems may be necessary to interpret application volumes, nozzle flow rates, and mixing directions for application equipment. The tables below provide the information to convert between U.S. units and metric units, and for calculating various dimensions for precision application.

VOLUME (LIQUID)

1 **gallon** (U.S.) = 4 quarts = 3.785 liters = 8 pints = 128 fluid ounces
1 **liter** = 0.264 U.S. gallon = 1.06 quarts = 2.1 pints = 1,000 milliliters
1 **U.S. gallon** = 0.83 Imperial gallon
1 **quart** = 0.25 gallon = 2 pints = 32 fluid ounces = 946 milliliters
1 **pint** = 1.25 gallon =2 cups = 16 fluid ounces = 473 milliliters
1 **fluid ounce** (U.S.) = 1.2 fluid ounce (Imperial) = 29.57 milliliters
1 **milliliter** or cubic centimeter = 0.035 fluid ounce
1 **cup** = 8 fluid ounces = 0.5 pint = 0.25 quart = 236.5 milliliters

WEIGHT

1 **gallon** (U.S.) of water = 8.34 pounds
1 **gallon** (Imperial) of water = 10.0 pounds
1 **pint** (U.S.) of water = 1.0 pound
1 **liter** of water = 1.0 kilogram
1 **milliliter** of water = 1.0 gram
1 **pound** = 0.45 kilograms = 16 ounces = 454 grams
1 **kilogram** = 2.2 pounds = 35.2 ounces = 1000 grams
1 **ounce** = 28.35 grams
1 **gram** = 0.0022 pounds = 0.035 ounces = 1000 milligrams
1 **pound of water** = 0.1198 gallons (U.S.)

LINEAR MEASURE

1 **inch** = 25.4 mm = 2.54 cm
1 **foot** = 30.48 cm
1 **yard** = 0.91 meter
1 **millimeter** = 0.03937 inch
1 **centimeter** = 0.3937 inch
1 **meter** = 3.281 feet
1 **kilometer** = 3,281 feet

CUBIC MEASURE

Cubic foot = 1,728 cubic inches = 0.037 cubic yard = 0.03 meter
Cubic foot = 7.48 gallons (U.S.) = 59.84 pints = 29.92 quarts = 28.32 liters
Ounces (liquid) = 1.805 cubic inches = 0.0625 pint = 0.03125 quart
Liter = 1,000 cubic centimeters = 0.035 cubic foot = 61.02 cubic inches
Cubic inches (cu in) → milliliter (ml) x 16.387
Cubic feet (cu ft) → cubic meter (cu m) x 0.0283
Cubic feet (cu ft) (volume) → liter (l) x 28.32
Cubic yards (cu yd) → cubic meter (cu m) x 0.7645

VELOCITY AND FLOW RATE

Feet per second (fps) → meter per second (m/s) x 0.3048
Feet per minute (fpm) → meter per second x 0.00508
Miles per hour (mph) → kilometer per hour (km/h) x 1.60934
Miles per hour (mph) → meter per second (m/s) x 0.44704
Gallons per minute (gpm) → liters per second (l/s) x 0.06309

AREA MEASURE

Square inches (sq in) → square centimeter (sq cm) x 6.4516
Square feet (sq ft) → square meter (sq m) x 0.0929
Square feet (sq ft) → linear feet ÷ 2.1
Square yards (sq yd) → square meter (sq m) x 0.8361
Square yards (sq yd) → square feet (sq ft) x 9.0

MISCELLANEOUS EQUIVALENTS

1 acre = 43,560 square feet = 4, 840 square yards
1 acre = 0.4047 hectare = 4,047 square meters
1 meter = 100 centimeters = 3.28 feet = 39.37 inches
1 foot = 12 inches = 30.48 centimeters
1 square foot = 2.1 linear feet
1 inch = 2.54 centimeters
2 miles per hour = 34 sec / 100 ft = 176 ft per minute
3 miles per hour = 23 sec / 100 ft = 264 ft per minute
4 miles per hour = 17 sec / 100 ft = 352 ft per minute

EQUIVALENT TABLES

°C	0	5	10	15	20	25	30	35	38	40
°F	32	41	50	59	68	77	86	95	100	104

Feet	1	2	3	4	5	6	7	8	9	10
Meter	0.30	0.60	0.91	1.21	1.52	1.82	2.13	2.43	2.74	3.04

Inches	1	2	3	4	5	6	7	8	9	10
Centimeters	2.54	5.08	7.62	10.1	12.7	15.2	17.7	20.3	22.8	25.4

Gallons	1	2	3	4	5	6	7	8	9	10
Liters	4.54	9.09	13.6	18.1	22.7	27.2	31.8	36.3	23.3	45.4

Cubic Feet	1	2	3	4	5	6	7	8	9	10
Cubic Meters	0.02	0.05	0.08	0.11	0.14	0.16	0.19	0.22	0.25	0.28

Pounds	1	2	3	4	5	6	7	8	9	10
Kilograms	0.45	0.90	1.36	1.81	2.26	2.72	3.17	3.62	4.08	4.53

Ounces	1	2	3	4	5	6	7	8	9	10
Milliliters	29.5	59.1	88.7	118.2	147.8	177.4	206.9	236.5	266.1	295.7

Square Feet	1	2	3	4	5	6	7	8	9	10
Square Meters	0.09	0.18	0.27	0.37	0.46	0.55	0.65	0.74	0.83	0.92

TEMPERATURE

Fahrenheit → Celcius = °F minus 32 then multiply by 0.5555
Celcius → Fahrenheit = °C plus 17.98 then multiply by 1.8

AREA AND VOLUME CALCULATION

Circle: Area = 3.14 x radius^2; Area = 0.7854 x diameter^2
Square or Rectangle: Area = length x width; Volume = length x width x height
Triangle: Area = 0.5 x base x height; Volume = ½ height x length x width
Cylinder: Volume = radius^2 x 3.1416 x height

PARTS PER MILLION AND PERCENTAGE CONCENTRATION

1 part per million (ppm) = 1 milligram / liter
= 1 milligram / kilogram
= 0.013 ounce in 100 gallons of water
= 0.0001 percent

1 percent = 10,000 ppm
= 10 grams per liter
= 10 grams per kilogram
= 1.33 oz by wt per gallon of water
= 8.34 pounds per 100 gallons of water

CONVERSION TABLE

	A	B	A→B	B→A
Weight	oz	g	x 28.35	x 0.0353
	lb	kg	x 0.454	x 2.205
Surface area	in^2	cm^2	x 6.45	x 0.155
	ft^2	m^2	x 0.093	x 10.764
	yd^2	m^2	x 0.836	x 1.196
Length	in	cm	x 2.54	x 0.394
	ft	m	x 0.305	x 3.281
	yd	m	x 0.914	x 1.094
	mile	km	x 1.609	x 0.621
Velocity	ft/s	m/s	x 0.305	x 3.281
	ft/min	m/s	x 0.00508	x 197.0
	mile/h	km/h	x 1.609	x 0.621
	mile/h	ft/min	x 88.0	x 0.0113
	m/s	km/h	x 3.61	x 0.277
Amount / area	lb/acre	kg/ha	x 1.12	x 0.894
	mg/ft^2	mg/m^2	x 10.794	x 0.093
	(US) gal/acre	liters/ha	x 9.346	x 0.107
	(US) fl oz/acre	ml/ha	x 73.14	x 0.0137
Volume	in^3	ft^3	x 0.000579	x 1728
	ft^3	yd^3	x 0.037	x 27
	yd^3	m^3	x 0.764	x 1.308
	ft^3	m^3	x 0.0283	x 35.31
	(US) fl oz	ml	x 29.6	x 0.0338
	(US) gal	liter	x 3.785	x 0.264
Pressure	lb/in^2	kg/cm	x 0.0703	x 14.22
	lb/in^2	bar	x 0.0689	x 14.504
	lb/in^2	kPa	x 6.89	x 0.145
	lb/m^2	atm	x 0.068	x 14.696

APPLICATION CALCULATIONS

Space Spraying. Calculating treatment volumes of storage rooms, dining areas of restaurants, or other large spaces is important in determining the amount of insecticide needed for space treatments. Some calculations are simple; for square and rectangular spaces, multiply the length times the width times the height. Complex building designs require dividing the space into geometric parts then determining the volume of each part, finally adding together the volumes for the total volume.

Example. A room has a length of 20 feet, it is 17 feet wide, the walls are 8 feet high, and the ceiling slopes to a peak in the center where the top ridge is 14 feet above the floor. The insecticide label directions for ultra-low volume (ULV) fogging require application of 10 ounces per 1000 cubic feet.

How many ounces will be needed to treat (space spray) the room?

Determine separately the volume of the rectangular space and triangular space above. Add these two to determine the volume of the entire room; then calculate the amount of liquid to be applied by a ULV fogger.

Rectangular room space: Volume = length x width x height
V = 20 feet x 17 feet x 8 feet
Volume = 2,720 cubic feet

Triangular ceiling space: Volume = length x width x height ÷ 2
V = 20 feet x 17 feet x 6 feet
Volume = 1,020 cubic feet

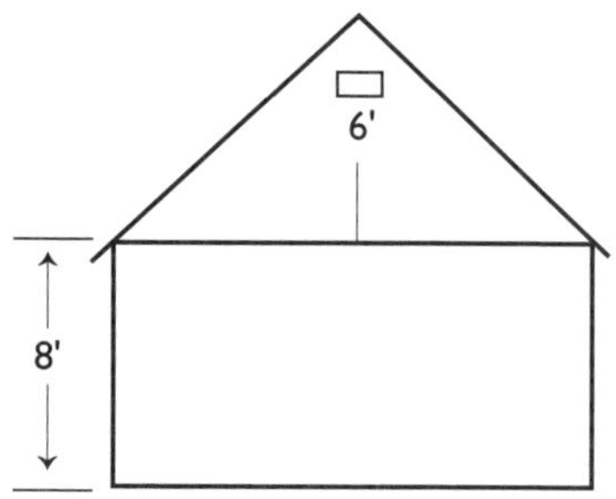

Total volume:

Rectangular space – 2,720 cubic feet
Triangular space – 1,020 cubic feet
= 3,740 cubic feet

Insecticide amount: 10 ounces / 1000 cubic feet = 37 ounces / 3,740 cubic feet

Space Treatment for Storage Rooms. Deductions must be made for machinery, furnishings, boxes, and other large objects in the room when calculating the volume for space treatments. For example, if a room contains 2,000 cubic feet of space, but approximately a quarter of this is taken up by equipment and stacked boxes, the amount of insecticide used should be based on 1,500 cubic feet.

The common objects in a room have a basic geometric shape and their volume can be approximated. Storage boxes, cabinets, refrigerators, and tables are square or rectangular. The volume of a cube or rectangular space is the length times the width times the height (L x W x H). The volume of a triangular space, such as a peaked roof, is ½ the height times the length times the width (1/2 H x L x W).

Example. A room has a length of 15 feet, a width of 30 feet, and the ceiling height is 8 feet. There are six large boxes along one wall, each box is 24 inches high, 24 inches wide, and 36 inches long. Along a wall is a floor cabinet that is 36 inches high, 24 inches wide, and 48 inches long; above this are three wall cabinets that are each 12 inches wide, 48 inches long, and 36 inches high. Insecticide label directions for ULV fogging require the application of 5 ounces per 1000 cubic feet.

How many ounces will be needed to treat (space spray) the room?

ROOM VOLUME:

Rectangular room space: Volume = length x width x height, V = 15 feet x 30 feet x 8 feet
Volume = 3,600 cubic feet

Volume of room contents: Floor cabinet = 24 cubic feet
Wall cabinets (3) = 36 cubic feet
Boxes (6) = 96 cubic feet
Total = 156 cubic feet

Total room volume (minus the contents) = 3,444 cubic feet
Insecticide amount: 5 ounces / 1000 cubic feet = 17 ounces / 3,444 cubic feet

Pre-Construction Termiticide Application. Pre-construction treatments include establishing a horizontal zone of treated soil that will be covered by a concrete slab. The typical treatment rate is 1 gallon per 10 square feet.

Example. A warehouse under construction has a slab-on-grade foundation that is 50 feet long and 30 feet wide. How many gallons of termiticide are required to treat this building, following the label specified rate of 1 gallon per 10 square feet?

Foundation: Square feet = length x width / Square feet = 50 x 30
Total = 1,500 sq ft

Insecticide amount: 1 gallon / 10 square feet = 150 gallons / 1,500 square feet

Post-Construction Termiticide Application. Post-construction treatments include all applications that are made after the final grading of the soil around the building. The applicator must treat by trenching and rodding into the trench or trench only along the foundation wall. The application rate is 4 gallons per 10 linear feet per each foot of depth.

Example. The exterior perimeter soil of a house will be treated with liquid termiticide. The foundation is angular, with linear feet dimensions of 50 feet, 40 feet, 30 feet, 20 feet, and 10 feet. The footing is 3 feet below grade.

How many gallons of termiticide are required to treat, following the specification of applying 4 gallons per 10 linear feet per foot of depth?

Foundation: Linear feet = 50 + 40 + 20 + 10 + 30 + 30
Total = 180 linear feet

Insecticide amount: 4 gallons per 10 linear feet per foot of depth
Total linear feet ÷ 10 x depth of foundation x 4 gallons
180 ÷ 10 = 18 x 3 = 54 x 4 = 216 gallons

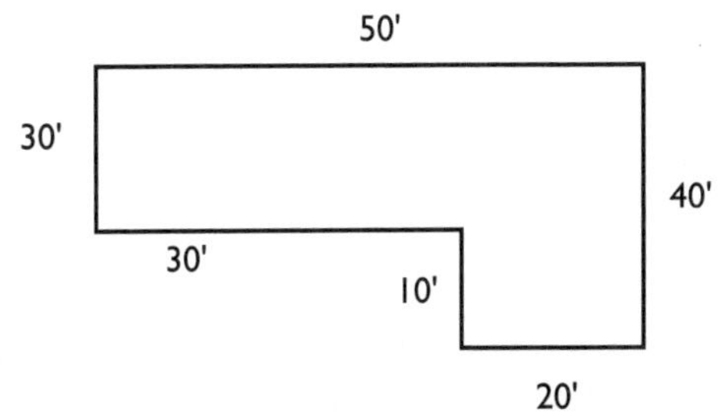

NOZZLE DELIVERY RATES

The operating pressure of a delivery system can have a significant effect on flow rate and quality of the spray distribution. Lowering the pressure reduces the spray angle and reduces flow rate. High pressure increases rate of flow through a nozzle. The practical application for spraying is: flow rate is reduced by 1/4 with a two-fold decrease in pressure, and reduced by 1/2 with a four-fold decrease in pressure. As pressure increases, droplet size increases, and the rate of nozzle wear tends to increase.

OUNCES/GALLONS PER MINUTE FLOW RATE FROM FAN NOZZLES

	OPERATING PRESSURE			
Fan Nozzle	20 psi	25 psi	30 psi	40 psi
50015 (coarse fan)	14 oz 0.11 gpm	15 oz 0.12 gpm	17 oz 0.13 gpm	19 oz 0.15 gpm
800067 (fine fan)	6 oz 0.05 gpm	7 oz 0.055 gpm	8 oz 0.06 gpm	9 oz 0.067 gpm
8002	18 oz 0.14 gpm	20 oz 0.16 gpm	22 oz 0.17 gpm	26 oz 0.20 gpm
8004	36 oz 0.28 gpm	41 oz 0.32 gpm	45 oz 0.35 gpm	51 oz 0.40 gpm
8006	54 oz 0.42 gpm	60 oz 0.47 gpm	66 oz 0.52 gpm	77 oz 0.60 gpm

CHAPTER 21

ADDITIONAL READING

Most of these articles are available from the websites of the industry magazines, Pest Control Technology [pctonline.com/magazine] and Pest Management Professional (originally Pest Control) [mypmp.net].

CHAPTER 1: THE NEXT GENERATION SERVICE TECHNICIAN

Thomas, C. R. and W. H Robinson – 1991. Technician training: current status and future trends. Pest Control Technology. (May): 78, 80-83.

CHAPTER 2: IPM: DETECTION AND MONITORING

Bruesch, J. – 2010. Learn to be organic. Pest Management Professional. (September): 57-61.

Cooper, R. – 2007. Four-legged bed bug detectives. Pest Control Technology. (August): 76-77.

Gulmahamad, H. – 1996. Getting moths "out of the mood." Pest Management Professional. (October): 60, 62, 63.

Jacobi, H. B. and W. H Robinson – 1997. Monitoring: The next generation. Pest Control Technology. (March): 34, 35, 38, 43, 110.

Mabbett, T. – 2005. Pheromones and IPM. Pest Management Professional. (May): 55, 56, 58, 60, 62.

Nansan, C. and T. Phillips. – 2004. Building a better pheromone trap. Pest Management Professional. (August): 66, 68.

Robinson, W. H – 1996. IPM in the Urban Ecosystem. Bulletin Entomological Society of America 42 (2): 76-78.

Robinson, W. H – 1995. Perspective on IPM: A critical review. Pest Control Technology. (June): 34-45.

Wang, C., T. J. Gibb, and G. W. Bennett – 2009. Interceptors assist in bed bug monitoring. Pest Control Technology. (April): 110.

CHAPTER 3: NON-CHEMICAL CONTROL METHODS

Ballard, J. B. – 2008. A new option (mattress liner for bed bug control). Pest Control Technology. (October): 144, 146, 147.

Farmer, B. R. and W. H Robinson – 1984. Is caulking beneficial for cockroach control? Pest Management Professional. (June): 28, 30, 32.

Getty, G. M., R. L. Taylor, and V. R. Lewis – 2008. Hot house: Can heat be used to control bed bugs and their eggs? Pest Control Technology. (February): 97-99.

Gold, R. E., H. N. Howell, and T. N. Decker – 1993. The cat's out of the bag. Pest Control Technology. (September): 40, 41, 45, 48, 109.

Potter, M. F., A. Romero, K. F. Haynes, and T. Jarzynka – 2008. Bed bugs, heat and hotel rooms. Pest Control Technology. (October): 106, 108, 109, 112, 114, 116, 118, 120, 121.

Quarles, W. – 2007. Diatomaceous earth. Pest Control Technology. (May): 64, 65, 66, 68.

Robinson, W. H – 2005. How it works: Bed bugs and bean leaves. Pest Management Professional. (April): 73.

CHAPTER 4: CHEMICAL INSECTICIDES

Yamamoto, I. and J. E. Casida, eds. – 1999. Nicotinoid Insecticides and the nicotinic acetycholine receptor. Springer-Verlag, Tokyo.

CHAPTER 5: INSECTICIDE FORMULATIONS

Bao, N. – 2007. Battling bait aversion. Pest Control. (September): 70-73.

Quarles, W. – 2007. Diatomaceous earth. Pest Control Technology. (May): 64-68.

CHAPTER 6: APPLICATION TECHNOLOGY

Barlow, R. and W. H Robinson – 1993. Insecticide application technology: Termiticide technology. Pest Control Technology. (May): 62, 66, 70, 72, 76.

Barlow, R. and W. H Robinson – 1993. Using foam effectively. Pest Control Technology. (November): 39, 42, 43, 44, 46.

Robinson, W. H – 1996. Application is the key to wood protection. Pest Management Professional. (October): 70-72.

Robinson, W. H – 2004. How it Works: The ABCs of ULV. Pest Management Professional. (September): 45, 48.

Robinson, W. H – 2005. How it Works: Flea control carpet treatment. Pest Management Professional. (August): 67.

Robinson, W. H – 2005. How it Works: Less droplets, same effectiveness on cockroaches. Pest Management Professional. (October): 114-115.

Robinson, W. H and R. Barlow – 1998. Is there risk of contamination? Pest Management Professional. (May): 30-34.

Zhai, J. and W. H Robinson – 1992. Insecticide application technology: Chemical control of the German cockroach. Pest Control Technology. (March): 40, 41, 44.

Zhai, J. and W. H Robinson – 1992. Insecticide application technology: Methods and equipment for cockroach control. Pest Control Technology. (June): 78-84.

CHAPTER 7: APPLICATION ECONOMICS

Ajjan, I., D. Zhang, and W. H Robinson – 1997. Cockroach baits. Pest Control Technology. (June): 47-56.

Byron, D.W. and W. H Robinson – 1988. Spraying correctly. Pest Management Professional. (May): 28, 30, 32, 39.

Greess, A. – 2008. Spray equipment cost: Purchase price tip of the iceberg. Pest Management Professional (June): 76.

Hinderer, C. – 2005. Care and feeding of equipment. Pest Management Professional. (August): 42, 44.

Robinson, W. H – 2010. Optimizing equipment performance to save money. Pest Control Technology. (October): 162, 164-168.

CHAPTER 8: TRANSFER OF A TOXIC DOSE

Robinson, W. H – 2005. How it Works: It's all about the legs. Pest Management Professional. (February): 58-59.

Suarez, M. and B. Thorne – 2000. Does every termite need a bite of the poison apple? Pest Management Professional. (February): 34, 35, 94.

Zhai, J. and W. H Robinson – 1990. The walking behavior of German cockroaches. Pest Control Technology. (July): 44-45.

CHAPTER 9: TERMITE CONTROL

Barlow, R. and W. H Robinson – 1993. Using foam effectively. Pest Control Technology. (November): 39, 42, 44, 46.

Forschler, B. – 1998. Termite baiting: A research perspective. Pest Control Technology. (June): 90, 92, 94, 96, 98.

Forschler, B. and V. Lewis – 1997. Why termites can dodge your treatment. Pest Management Professional. (February): 42-46, 53.

Long, C. E., B. L. Thorne, and N. L. Breisch – 2001. Termites and mulch. Pest Control Technology. (October): 64, 66, 68, 70.

Mampe, D. and B. Brett – 1992. Rodding, soil type and application tips. Pest Management Professional (January): 48-50.

Robinson, W. H – 2004. Soil treatments, revisited. Pest Management Professional. (July): 47-49.

Robinson, W. H and P. Powell – 1990. Pre-treat facts. Pest Management Professional. (December): 32-34.

Sheperdigan, M. – 2010. What's under the slab? Pest Management Professional. (July): 38, 39.

Thomas, C. R. and W. H Robinson – 1991. Foam formulations for sub-slab injection of termiticides. Pest Control Technology. (November): 38, 42, 43.

Thomas, C. R. and W. H Robinson – 1991. Termiticide distribution after sub-slab application. Pest Control Technology. (October): 60, 64, 66.

CHAPTER 10: WOOD-INFESTING INSECT CONTROL

Barlow, R. and W. H Robinson – 1995. Weathering the elements: Boron insecticides. Pest Control Technology. (December): 46, 48, 50.

Green, A. – 2006. The .45 caliber bee. Pest Control Technology. (March): 82-84, 86, 88.

Green, A. – 2006. An all-American insect. Pest Control Technology. (August): 54, 55, 58, 130.

Powell, P. and W. H Robinson – 1989. Chemical control of wood-infesting beetles. Pest Control Technology. (February): 74, 76, 81.

Powell, P. and W. H Robinson – 1990. Wood injection system for the control of wood-infesting beetles. Pest Control Technology. (October): 34-35.

Powell, P. and W. H Robinson – 1990. Wood-infesting beetle control answers. Pest Management Professional. (April): 66-68.

Robinson, W. H and R. Barlow – 1995. Penetration of borates into structural wood. Pest Control Technology. (November): 56, 57, 58.

Robinson, W. H and R.A. Barlow – 1996. Penetration of boron into structural wood. Pest Control Technology. (January): 72, 73, 87.

Robinson, W. H, R. Barlow, and G. Rambo – 1996. How deep does it go? Pest Control Technology. (May): 66, 68.

CHAPTER 10: INDOOR INSECTICIDE APPLICATION

Byron, D. and W. H Robinson – 1987. Doing fleas correctly. Pest Management Professional. (April): 44-46, 48.

Byron, D. and W. H Robinson – 1988. Locating flea hot spots. Pest Management Professional. (June): 28-30.

Klotz, J., G. Taniguichi, H. Field, and K. Kupfer, 2010. Ant baiting. Pest Management Professional. (April): 40, 43, 44, 45.

CHAPTER 11: OUTDOOR INSECTICIDE APPLICATION

Barlow, R. and W. H Robinson – 1997. American cockroaches can be a finicky breed. Pest Management Professional. (July): 66-68.

Robinson, W. H – 2004. Perimeter primer. Pest Management Professional. (June): 49-50.

Yoder, K. and W. H Robinson – 1991. Biology and control of boxelder bugs. Pest Control Technology. (June): 30-31.

CHAPTER 13: RODENT CONTROL

Berman, R. and E. Smith – 2009. Using blacklight for a rodent inspection. Pest Control Technology. (January): 103-104.

Corrigan, B. – 2007. Frequently asked questions about rodenticide baits. Pest Control Technology. (October): 134-136.

Corrigan, B. – 2007. Rodent control in schools and other child-care facilities. Pest Control Technology. August): 42, 44, 46.

Corrigan, B. – 2008. The horizontal and vertical movements of city rodents. Pest Control Technology. (September): 127, 130.

Corrigan, B. – 2009. Bait station reactions. Pest Control Technology. (June): 105-107.

Corrigan, B. – 2010. Know when to strike. Pest Management Professional. (August): 25, 28, 30, 32, 34, 36, 38.

Greene, A. – 2004. The tooth of the rat. Pest Control Technology. (October): 81, 84, 86, 87.

Martin, P. – 2007. Voles or moles. Pest Management Professional. (October): 174-178.

Mitchell, S. – 2007. House mouse menace, a new medical perspective. Pest Control Technology. (August): 29-30.

CHAPTER 15: STRUCTURAL WOOD AND URBAN SOIL

Robinson, W. H – 2005. Separate wood-spraying fact from fiction. Pest Management Professional. (June): 64.

Yen, P.Y. – 1995. Water movement in soil. Pest Control Technology. (April): 86, 90, 92, 113.

INDEX

C

G

H

I

L

M

S

T